BIOZONE

Biology Modular W...

D1138300

University of Worcester, Peirson Building
Henwick Grove, Worcester. WR2 6AJ
Telephone: 01905 855341

Human Evolution

The Biozone Writing Team:

Tracey Greenwood

Lyn Shepherd

Richard Allan

Daniel Butler

Published by:
Biozone International Ltd
109 Cambridge Road, Hamilton 3216, New Zealand

Printed by REPLIKA PRESS PVT LTD

Distribution Offices:

United Kingdom & Europe	**Biozone Learning Media (UK) Ltd**, Scotland
	Telephone: +44 (131) 557 5060
	Fax: +44 (131) 557 5030
	Email: sales@biozone.co.uk
	Website: www.biozone.co.uk
USA, Canada, South America, Africa	**Biozone International Ltd**, New Zealand
	Telephone: +64 (7) 856 8104
	Freefax: 1-800717-8751 (USA-Canada only)
	Fax: +64 (7) 856 9243
	Email: sales@biozone.co.nz
	Website: www.biozone.co.nz
Asia & Australia	**Biozone Learning Media Australia**, Australia
	Telephone: +61 (7) 5575 4615
	Fax: +61 (7) 5572 0161
	Email: sales@biozone.com.au
	Website: www.biozone.com.au

© 2006 **Biozone International Ltd**
First Edition 2006
ISBN: 1-877329-89-4

Front cover photographs:

"Stag and Reindeer", Lascaux, Dordogne, France. Image from "Art of Antiquity",

Corel Corporation, Professional Photos

Australopithecus boisei KNM OH 5, reconstruction of skull with permission from
Skulls Unlimited, http://www.skullsunlimited.com/

Biology Modular Workbook Series

The Biozone *Biology Modular Workbook Series* has been developed to meet the demands of customers with the requirement for a modular resource which can be used in a flexible way. Like Biozone's popular Student Resource and Activity Manuals, these workbooks provide a collection of visually interesting and accessible activities, which cater for students with a wide range of abilities and background. The workbooks are divided into a series of chapters, each comprising an introductory section with detailed learning objectives and useful resources, and a series of write-on activities ranging from paper practicals and data handling exercises, to questions requiring short essay style answers. Material for these workbooks has been drawn from Biozone's popular, widely used manuals, but the workbooks have been structured with greater ease of use and flexibility in mind. During the development of this series, we have taken the opportunity to improve the design and content, while retaining the basic philosophy of a student-friendly resource which spans the gulf between textbook and study guide. With its unique, highly visual presentation, it is possible to engage and challenge students, increase their motivation and empower them to take control of their learning.

Human Evolution

This title in the *Biology Modular Workbook Series* provides students with a set of comprehensive guidelines and highly visual worksheets through which to explore aspects of hominin evolution. *Human Evolution* is the ideal companion for students in biology and anthropology, encompassing our position as primates, as well as the nature of human physical and cultural evolution. This workbook comprises three chapters each focussing on one particular area within this broad topic. These areas are explained through a series of activities, usually of one or two pages, each of which explores a specific concept (e.g. primate evolution or the development of intelligence). Model answers (on CD-ROM) accompany each order free of charge. *Human Evolution* is a student-centred resource and is part of a larger package, which also includes the **Human Evolution Presentation Media CD-ROM**. Students completing the activities, in concert with their other classroom and practical work, will consolidate existing knowledge and develop and practise skills that they will use throughout their course. This workbook may be used in the classroom or at home as a supplement to a standard textbook. Some activities are introductory in nature, while others may be used to consolidate and test concepts already covered by other means. Biozone has a commitment to produce a cost-effective, high quality resource, which acts as a student's companion throughout their biology study. Please do not photocopy from this workbook; we cannot afford to provide single copies of workbooks to schools and continue to develop, update, and improve the material they contain.

Acknowledgements and Photo Credits

Royalty free images, purchased by Biozone International Ltd, are used throughout this manual and have been obtained from the following sources: istockphotos (www.istockphoto.com) • Corel Corporation from various titles in their Professional Photos CD-ROM collection including *Art of Antiquity* and *Apes*; ©Hemera Technologies Inc, 1997-2001; © 2005 JupiterImages Corporation www.clipart.com; PhotoDisc®, Inc. USA, www.photodisc.com. Biozone's authors also acknowledge the generosity of those who have kindly provided photographs for this edition: • Dept of Biological Sciences, University of Waikato, for access to their collection of hominin skulls • David Haring at the Duke Lemur Center for the photograph of the aye aye • Skulls Unlimited International: www.skullsunlimited.com • Grotte de Rouffignac, for drawings and photographs of the Rouffignac Cave • Jan Morrison, for her drawings • The Melbourne Zoo for their assistance in photographing their splendid collection of primates. Coded credits are: **DA**: Donna Allan, **RA**: Richard Allan,

Also in this series:

Skills in Biology

Cell Biology & Biochemistry

Genes & Inheritance

Evolution

For other titles in this series go to:
www.thebiozone.com/modular.html

Contents

The Primates

Cultural Evolution

Hominin Evolution

Activity is marked: ▪ to be done; ✔ when completed

How to Use this Workbook

Human Evolution is designed to provide students with a resource that will make the acquisition of knowledge and skills in this exciting and controversial area easier and more enjoyable. An understanding of our human origins is essential for students of anthropology and an important component of many standard biology courses. Moreover, this subject is of high interest, with many competing ideas based of differing interpretations of the fossil record. This workbook is suitable for all students; it is both thorough and engaging, and provides ample opportunity for students to consolidate and extend their knowledge in this area. It is **not a textbook**; its aim is to complement the texts written for your particular course. *Human Evolution* provides the following resources in each chapter. You should refer back to them as you work through each set of worksheets.

Guidance Provided for Each Topic

Learning objectives:

These provide you with a map of the chapter content. Completing the learning objectives relevant to your course will help you to satisfy the knowledge requirements of your syllabus. Your teacher may decide to leave out points or add to this list.

Chapter content:

The upper panel of the header identifies the general content of the chapter. The lower panel provides a brief summary of the chapter content.

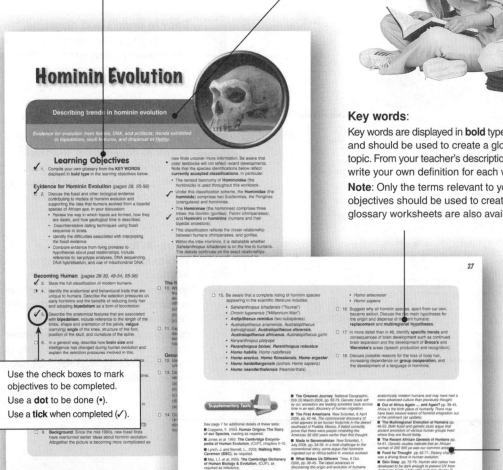

Key words:

Key words are displayed in **bold** type in the learning objectives and should be used to create a glossary as you study each topic. From your teacher's descriptions and your own reading, write your own definition for each word.

Note: Only the terms relevant to your selected learning objectives should be used to create your glossary. Free glossary worksheets are also available from our web site.

Use the check boxes to mark objectives to be completed.
Use a **dot** to be done (•).
Use a **tick** when completed (✓).

Supplementary texts:

References to supplementary texts suitable for use with this workbook are provided. Chapter references are provided as appropriate. The details of these are provided on page 7, together with other resources information.

Supplementary resources

Biozone's Presentation MEDIA are noted where appropriate.

Periodical articles:

Ideal for those seeking more depth or the latest research on a specific topic. Articles are sorted according to their suitability for student or teacher reference. Visit your school, public, or university library for these articles.

Internet addresses:

Access our database of links to more than **800** web sites (updated regularly) relevant to the topics covered. Go to Biozone's own web site: **www.thebiozone.com** and link directly to listed sites using the *BioLinks* button.

Activity Pages

The activities and exercises make up most of the content of this workbook. They are designed to reinforce the concepts you have learned about in the topic. Your teacher may use the activity pages to introduce a topic for the first time, or you may use them to revise ideas already covered. They are excellent for use in the classroom, and as homework exercises and revision. In most cases, the activities should not be attempted until you have carried out the necessary background reading from your textbook. As a self-check, model answers for each activity are provided on CD-ROM with each order of workbooks.

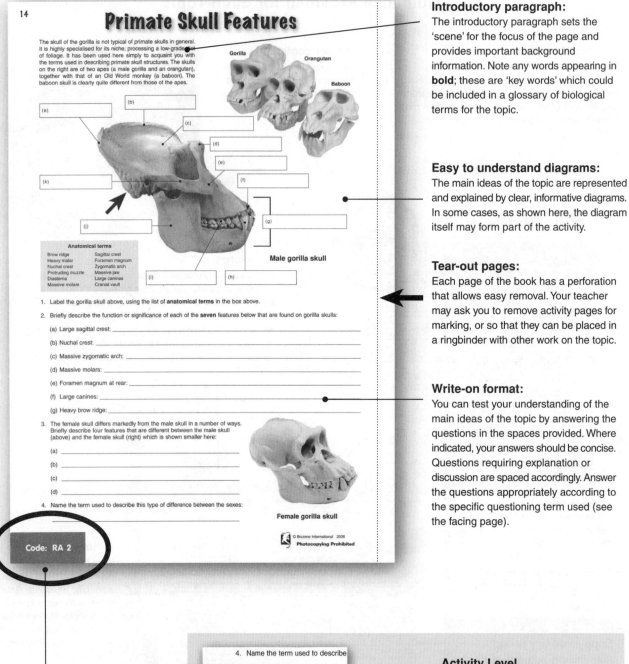

Introductory paragraph:
The introductory paragraph sets the 'scene' for the focus of the page and provides important background information. Note any words appearing in **bold**; these are 'key words' which could be included in a glossary of biological terms for the topic.

Easy to understand diagrams:
The main ideas of the topic are represented and explained by clear, informative diagrams. In some cases, as shown here, the diagram itself may form part of the activity.

Tear-out pages:
Each page of the book has a perforation that allows easy removal. Your teacher may ask you to remove activity pages for marking, or so that they can be placed in a ringbinder with other work on the topic.

Write-on format:
You can test your understanding of the main ideas of the topic by answering the questions in the spaces provided. Where indicated, your answers should be concise. Questions requiring explanation or discussion are spaced accordingly. Answer the questions appropriately according to the specific questioning term used (see the facing page).

Activity code:
Activities are coded to help you in identifying the type of activities and the skills they require. Most activities require some basic knowledge recall, but will usually build on this to include applying the knowledge to explain observations or predict outcomes. The least difficult questions generally occur early in the activity, with more challenging questions towards the end of the activity.

> 4. Name the term used to describe
>
> **Code: RA 2**
>
> * Material to assist with the activity may be found on other pages of the workbook or in textbooks.

Activity Level
1 = Simple questions not requiring complex reasoning
2 = Some complex reasoning may be required
3 = More challenging, requiring integration of concepts

Type of Activity
D = Includes some data handling and/or interpretation
P = Includes a paper practical
R = May require research outside the information on the page, depending on your knowledge base*
A = Includes application of knowledge to solve a problem
E = Extension material

Explanation of Terms

Questions come in a variety of forms. Whether you are studying for an exam or writing an essay, it is important to understand exactly what the question is asking. A question has two parts to it: one part of the question will provide you with information, the second part of the question will provide you with instructions as to how to answer the question. Following these instructions is most important. Often students in examinations know the material but fail to follow instructions and do not answer the question appropriately. Examiners often use certain key words to introduce questions. Look out for them and be clear as to what they mean. Below is a description of terms commonly used when asking questions in biology.

Commonly used Terms in Biology

The following terms are frequently used when asking questions in examinations and assessments. Students should have a clear understanding of each of the following terms and use this understanding to answer questions appropriately.

Account for: Provide a satisfactory explanation or reason for an observation.

Analyse: Interpret data to reach stated conclusions.

Annotate: Add **brief** notes to a diagram, drawing or graph.

Apply: Use an idea, equation, principle, theory, or law in a new situation.

Appreciate: To understand the meaning or relevance of a particular situation.

Calculate: Find an answer using mathematical methods. Show the working unless instructed not to.

Compare: Give an account of similarities and differences between two or more items, referring to both (or all) of them throughout. Comparisons can be given using a table. Comparisons generally ask for similarities more than differences (see contrast).

Construct: Represent or develop in graphical form.

Contrast: Show differences. Set in opposition.

Deduce: Reach a conclusion from information given.

Define: Give the precise meaning of a word or phrase as concisely as possible.

Derive: Manipulate a mathematical equation to give a new equation or result.

Describe: Give a detailed account, including all the relevant information.

Design: Produce a plan, object, simulation or model.

Determine: Find the only possible answer.

Discuss: Give an account including, where possible, a range of arguments, assessments of the relative importance of various factors, or comparison of alternative hypotheses.

Distinguish: Give the difference(s) between two or more different items.

Draw: Represent by means of pencil lines. Add labels unless told not to do so.

Estimate: Find an approximate value for an unknown quantity, based on the information provided and application of scientific knowledge.

Evaluate: Assess the implications and limitations.

Explain: Give a clear account including causes, reasons, or mechanisms.

Identify: Find an answer from a number of possibilities.

Illustrate: Give concrete examples. Explain clearly by using comparisons or examples.

Interpret: Comment upon, give examples, describe relationships. Describe, then evaluate.

List: Give a sequence of names or other brief answers with no elaboration. Each one should be clearly distinguishable from the others.

Measure: Find a value for a quantity.

Outline: Give a brief account or summary. Include essential information only.

Predict: Give an expected result.

Solve: Obtain an answer using algebraic and/or numerical methods.

State: Give a specific name, value, or other answer. No supporting argument or calculation is necessary.

Suggest: Propose a hypothesis or other possible explanation.

Summarise: Give a brief, condensed account. Include conclusions and avoid unnecessary details.

In Conclusion

Students should familiarise themselves with this list of terms and, where necessary throughout the course, they should refer back to them when answering questions. The list of terms mentioned above is not exhaustive and students should compare this list with past examination papers / essays etc. and add any new terms (and their meaning) to the list above. The aim is to become familiar with interpreting the question and answering it appropriately.

Using the Internet

The internet is a vast global network of computers connected by a system that allows information to be passed through telephone connections. When people talk about the internet they usually mean the **World Wide Web** (WWW). The WWW is a service that has made the internet so simple to use that virtually anyone can find their way around, exchange messages, search libraries and perform all manner of tasks. The internet is a powerful resource for locating information. Listed below are two journal articles worth reading. They contain useful information on what the internet is, how to get started, examples of useful web sites, and how to search the internet.

- **Click Here: Biology on the Internet** Biol. Sci. Rev., 10(2) November 1997, pp. 26-29.
- **An A-level biologists guide to The World Wide Web** Biol. Sci. Rev., 10(4) March 1998, pp. 26-29.

Using the Biozone Website: www.thebiozone.com

The **Back** and **Forward** buttons allow you to navigate between pages displayed on a www site

The current **internet address (URL)** for the web site is displayed here. You can type in a new address directly into this space.

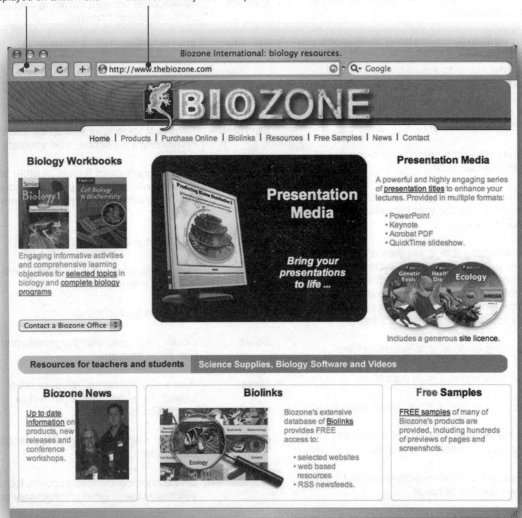

Searching the Net

The WWW addresses listed throughout this workbook have been selected for their relevance to the topic in which they are listed. We believe they are good sites. Don't just rely on the sites that we have listed. Use the powerful 'search engines', which can scan the millions of sites for useful information. Here are some good ones to try:

Alta Vista:	www.altavista.com
Ask Jeeves:	www.ask.com
Excite:	www.excite.com/search
Google:	www.google.com
Go.com:	www.go.com
Lycos:	www.lycos.com
Metacrawler:	www.metacrawler.com
Yahoo:	www.yahoo.com

Biozone International provides a service on its web site that links to all internet sites listed in this workbook. Our web site also provides regular updates with new sites listed as they come to our notice and defunct sites deleted. Our **BIO LINKS** page, shown below, will take you to a database of regularly updated links to more than 800 other quality biology web sites.

The **Resource Hub**, accessed via the homepage or resources, provides links to the supporting resources referenced in the workbook. These resources include comprehensive and supplementary texts, biology dictionaries, computer software, videos, and science supplies. These can be used to enhance your learning experience.

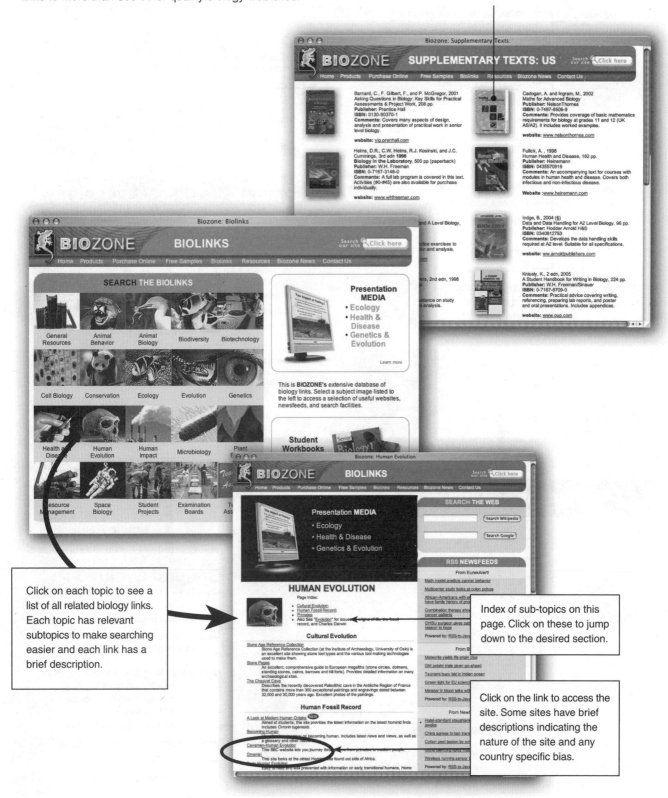

Click on each topic to see a list of all related biology links. Each topic has relevant subtopics to make searching easier and each link has a brief description.

Index of sub-topics on this page. Click on these to jump down to the desired section.

Click on the link to access the site. Some sites have brief descriptions indicating the nature of the site and any country specific bias.

Concept Map for Human Evolution

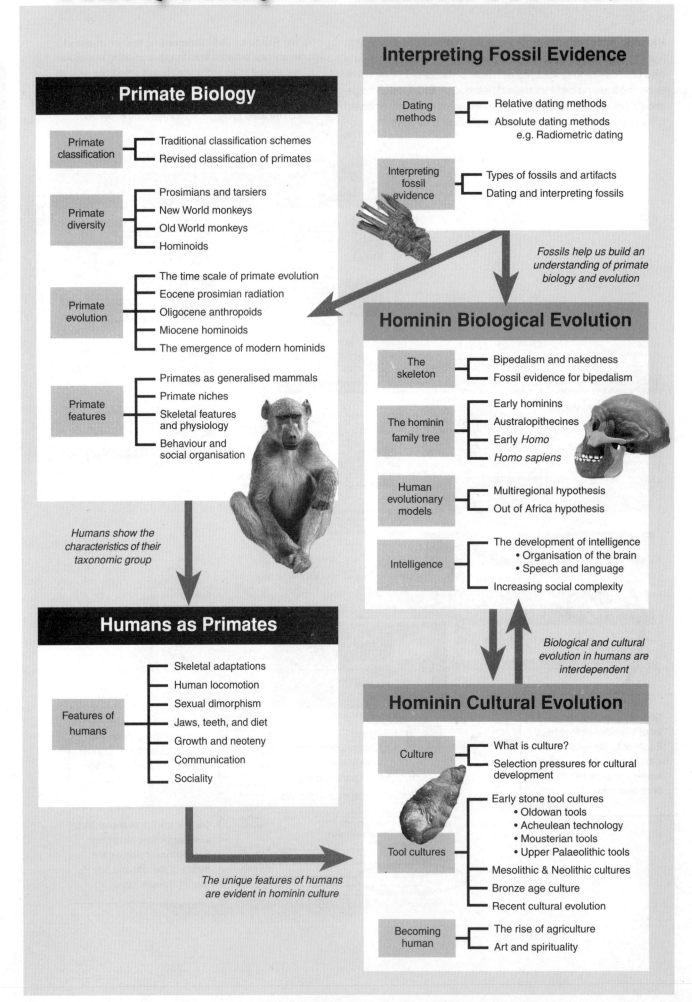

Primate Biology

Primate classification
- Traditional classification schemes
- Revised classification of primates

Primate diversity
- Prosimians and tarsiers
- New World monkeys
- Old World monkeys
- Hominoids

Primate evolution
- The time scale of primate evolution
- Eocene prosimian radiation
- Oligocene anthropoids
- Miocene hominoids
- The emergence of modern hominids

Primate features
- Primates as generalised mammals
- Primate niches
- Skeletal features and physiology
- Behaviour and social organisation

Interpreting Fossil Evidence

Dating methods
- Relative dating methods
- Absolute dating methods e.g. Radiometric dating

Interpreting fossil evidence
- Types of fossils and artifacts
- Dating and interpreting fossils

Fossils help us build an understanding of primate biology and evolution

Hominin Biological Evolution

The skeleton
- Bipedalism and nakedness
- Fossil evidence for bipedalism

The hominin family tree
- Early hominins
- Australopithecines
- Early *Homo*
- *Homo sapiens*

Human evolutionary models
- Multiregional hypothesis
- Out of Africa hypothesis

Intelligence
- The development of intelligence
  - Organisation of the brain
  - Speech and language
- Increasing social complexity

Humans show the characteristics of their taxonomic group

Humans as Primates

Features of humans
- Skeletal adaptations
- Human locomotion
- Sexual dimorphism
- Jaws, teeth, and diet
- Growth and neoteny
- Communication
- Sociality

Biological and cultural evolution in humans are interdependent

Hominin Cultural Evolution

Culture
- What is culture?
- Selection pressures for cultural development

Tool cultures
- Early stone tool cultures
  - Oldowan tools
  - Acheulean technology
  - Mousterian tools
  - Upper Palaeolithic tools
- Mesolithic & Neolithic cultures
- Bronze age culture
- Recent cultural evolution

Becoming human
- The rise of agriculture
- Art and spirituality

The unique features of humans are evident in hominin culture

Resources Information

Your set textbook should always be a starting point for information, but there are also many other resources available. A list of readily available resources is provided below. Access to the publishers of these resources can be made directly from Biozone's web site through our resources hub: **www.thebiozone.com/resource-hub.html**. Please note that our listing of any product in this workbook does not denote Biozone's endorsement of it.

Supplementary Texts

The literature base for this topic is immense and often aimed at specialist readers. The titles listed here have been chosen on the basis of their accuracy, up-to-date content, and accessibility and appeal to students.

Coppens, Y., 2004 (English language edn)
Human Origins: The Story of Our Species, 180 pp. **Publisher**: Hachette Illustrated UK, Octopus Publishing Group Ltd
ISBN: 1-84430-095-1
Comments: *An appealing but informative read based on reconstructions for film of the history of human physical and cultural evolution.*

Dunbar, R. and L. Barrett, 2000
Cousins: Our Primate Relatives, 240 pp.
Publisher: BBC Worldwide Ltd
ISBN: 0-563-55115-1
Comments: *An excellent resource for studying primate origins, and the physical and behavioural features of the primate order. Well organised and superbly illustrated.*

Jones, S., R. Martin, and D. Pilbeam (eds), 1992
The Cambridge Encyclopedia of Human Evolution, 506 pp. including appendices and index
Publisher: Cambridge University Press
ISBN: 0-521-46786-1
Comments: *Primarily a teacher's reference, which includes material on primates as well as full coverage of hominin evolution.*

Lynch, J. and L. Barrett, 2002
Walking With Cavemen, 224 pp.
Publisher: Headline Book Publishing
ISBN: 0-7553-1177-9
Comments: *The story of hominin evolution, presented as an appealing student read and illustrated throughout with reconstructions based on the film of the same name.*

Mai, L.L., M. Young Owl. and M.P. Kersting, 2005
The Cambridge Dictionary of Human Biology and Evolution, 650 pp. including appendices
Publisher: Cambridge University Press
ISBN: 0-521-66486-1
Comments: *More encyclopaedia than dictionary, this is an excellent, authoritative, and up-to-date reference with a number of useful appendices.*

Rowe, N. 1996
The Pictorial Guide to the Living Primates, 263 pp. including index
Publisher: Pogonius Press
ISBN: 0-9648825-1-5
Comments: *More than a simple guide, this comprehensive text covers the entire primate order, with information on physical features, behaviour, sociality, and conservation status.*

Biology Dictionaries

Access to a good biology dictionary is useful when dealing with biological terms. Some of the titles available are listed below. Link to the relevant publisher via Biozone's resources hub or by typing: **www.thebiozone.com/resources/dictionaries-pg1.html**

Clamp, A. **AS/A-Level Biology. Essential Word Dictionary**, 2000, 161 pp. Philip Allan Updates.
ISBN: 0-86003-372-4.
Carefully selected essential words for AS and A2. Concise definitions are supported by further explanation and illustrations where required.

Hale, W.G., J.P. Margham, & V.A. Saunders.
Collins: Dictionary of Biology 3 ed. 2003, 672 pp. HarperCollins. **ISBN**: 0-00-714709-0.
Updated to take in the latest developments in biology from the Human Genome Project to advancements in cloning (new edition pending).

Henderson, I.F, W.D. Henderson, and E. Lawrence.
Henderson's Dictionary of Biological Terms, 1999, 736 pp. Prentice Hall. **ISBN**: 0582414989
This edition has been updated, rewritten for clarity, and reorganised for ease of use. An essential reference and the dictionary of choice for many.

McGraw-Hill (ed). **McGraw-Hill Dictionary of Bioscience**, 2 ed., 2002, 662 pp. McGraw-Hill.
ISBN: 0-07-141043-0
22 000 entries encompassing more than 20 areas of the life sciences. It includes synonyms, acronyms, abbreviations, and pronunciations for all terms.

Periodicals, Magazines, & Journals

Biological Sciences Review: *An informative quarterly publication for biology students.* Enquiries: **UK**: Philip Allan Publishers **Tel**: 01869 338652 **Fax**: 01869 338803 **E-mail**: sales@philipallan.co.uk **Australasia**: **Tel**: 08 8278 5916, **E-mail**: rjmorton@adelaide.on.net

New Scientist: *Widely available weekly magazine with research summaries and features.* Enquiries: Reed Business Information Ltd, 51 Wardour St. London WIV 4BN **Tel**: (UK and intl):+44 (0) 1444 475636 **E-mail**: ns.subs@qss-uk.com *or subscribe from their web site.*

Scientific American: *A monthly magazine containing specialist features. Articles range in level of reading difficulty and assumed knowledge.* Subscription enquiries: 415 Madison Ave. New York. NY10017-1111 **Tel**: (outside North America): 515-247-7631 **Tel**: (US & Canada): 800-333-1199

School Science Review: *A quarterly journal which includes articles, reviews, and news on current research and curriculum development. Free to Ordinary Members of the ASE or available on subscription.* Enquiries: **Tel**: 01707 28300 **Email**: info@ase.org.uk *or visit their web site.*

The American Biology Teacher: *The peer-reviewed journal of the NABT. Published nine times a year and containing information and activities relevant to biology teachers.* Contact: NABT, 12030 Sunrise Valley Drive, #110, Reston, VA 20191-3409 **Web**: www.nabt.org

The Primates

Understanding primate legacy in the evolution of humans

Humans as primates, comparing humans and other living hominoids, primate classification, primate characteristics, primate origins

Learning Objectives

☐ 1. Compile your own glossary from the **KEY WORDS** displayed in **bold type** in the learning objectives below.

Primate Classification *(pages 9-12, 15-16)*

☐ 2. Using modern revisions of primate classification, define the terms: **primate, prosimian, tarsier, anthropoid, hominoid**. Giving examples, describe the characteristics that all primates have in common.

☐ 3. Recognise taxonomic groupings under the traditional scheme for primate classification, particularly the **lesser apes** (hylobatids) and **great apes** (pongids). Appreciate that this traditional classification does not accurately reflect the true phylogeny of the apes.

☐ 4. Describe the current classification of the Order Primates recognising the following: **prosimians, tarsiers, New World monkeys, Old World monkeys, hylobatids, hominids,** and **hominins**. Appreciate that the revised classification of **hominoids** accommodates our current understanding of the genetic similarities and differences between the **apes** and humans.

Primate Characteristics *(pages 9-14, 17-22)*

☐ 5. Clearly understand what is meant by a **distinguishing characteristic** (feature). List the distinguishing physical characteristics of different primate groups. Include reference to each of the following:
- Facial structure and dentition.
- Snout and nostril structure.
- Development of olfactory organs.
- Limb structure and tail function (**prehensile** or not).
- Extent of bifocal and colour vision.

☐ 6. List the distinguishing **behavioural** characteristics of different primate groups. Describe each of the following as appropriate: social behaviour, communication, locomotion, behaviour towards young.

Primate Origins *(pages 15-16, 23-25)*

☐ 7. Describe the likely origin of the primates from a **common ancestor** about 60 million years ago.

☐ 8. Explain how the various groups of primates have adapted to different habitats. Identify the different diets and modes of locomotion associated with the various niches occupied by primates.

☐ 9. Describe the evolution of the **hominoids** (monkeys and apes) during the Miocene-Early Pleistocene periods.

☐ 10. Describe the characteristics of the early ancestral forms that gave rise to the hominoids: **dryomorphs** (e.g. *Dryopithecus*, **Proconsul**) and **ramamorphs** (e.g. *Sivapithecus, Gigantopithecus, Ouranopithecus*).

See page 7 for additional details of these texts:

■ Dunbar, R. and L. Barrett, 2000. **Cousins** BBC Worldwide Ltd., as reqd.

■ Jones *et al.* 1992. **The Cambridge Encyclopedia of Human Evolution**, (CUP), chapters 1-8.

■ Mai, L.I. *et al*, 2005. **The Cambridge Dictionary of Human Biology & Evolution**, (CUP), as reqd.

■ Rowe, N. 1996. **The Pictorial Guide to the Living Primates**, (Pogonias Press), as reqd.

See page 7 for details of publishers of periodicals:

■ **A Curious Kinship: Apes and Humans** National Geographic, 181(3) March 1992, pp. 2-53. *An examination of our closest relatives, the great apes. Includes discussion and images of the apes, and discusses their shared ancestry with humans.*

■ **The Greatest Apes** New Scientist, 15 May 1999, pp. 26-30. *An examination of the genetic similarities and differences between chimps and humans (includes chromosomal arrangements).*

■ **The Culture of Chimpanzees** Sci. American, Jan. 2001, pp. 48-55. *The social organisation of chimps allows opportunity for extensive learning experiences, and these may vary depending on the social context. This article includes a table of behaviours seen in different chimp communities.*

■ **Disturbing Behaviours of the Orangutan** Sci. American, June 2002, pp. 46-51. *Studies of orangs show that subordinate males remain in a state of arrested development, but copulate forcibly with females. This disturbing behaviour does not represent stress, but an evolutionary strategy aimed at maximising reproductive success.*

■ **An Ancestor to Call our Own** Sci. American, Jan. 2003, pp. 42-51. *Controversial new fossils (Sahelanthropus, Orririn, Ardipithecus) may be close to origin of humanity.*

BECOMING HUMAN Scientific American **Special Issue** 16(2) 2006. *A collection of articles (some updated revisions) covering aspects of primate behaviour and evolution. Includes:*

■ **Planet of the Apes** pp. 4-13. *The two ancient Eurasian ape lineages, one represented by Sivapithecus, the probable forebear of orangs, and the other by Dryopithecus, a possible ancestor of African apes and humans.*

■ **Bonobo Sex and Society** pp. 14-21. *Sexual behaviour maintains cohesion in these chimps.*

■ **Diet and Primate Evolution** pp. 22-29. *From an evolutionary viewpoint, we are what we ate. The first primates evolved in the canopy of the forests that proliferated during the late Cretaceous.*

■ **Why are Some Animals so Smart** pp. 48-55. *An unexpected finding of tool use in orangutans swamp suggests an answer to how and why intelligence develops in some species.*

See pages 4-5 for details of how to access **Bio Links** from our web site: **www.thebiozone.com** From Bio Links, access sites under the topics: **HUMAN EVOLUTION:** > **Primates:** • About primates • Bonobo sex and society • Chimpanzee hunting habits • Fossil primates 1 • Fossil primates 2 • Order primates * Physical anthropology tutorials • Primate images

Presentation MEDIA to support this topic:

HUMAN EVOLUTION

Prosimians and Tarsiers

Prosimians, the common name for the suborder Prosimii, means 'before the apes'. The group (now reclassified as the suborder **Strepsirhini**) includes the lemurs, lorises, pottos, and bushbabies. The lorises are found in southern Asia, the pottos and bush babies in Africa, while all members of the superfamily Lemuroidea (the lemurs, indrids and aye-aye) are found on the island of Madagascar. Prosimians are considered to be more primitive than other primates because some of their anatomical characteristics are found in some other mammals but not in anthropoid primates (see list of common features below, left).

The nocturnal, tree-dwelling **tarsiers**, which were once classified as prosimians, exhibit characteristics of both prosimians and anthropoids (monkeys, apes, and humans), while maintaining characteristics unique to themselves. They are now classified in the suborder **Haplorhini** as intermediate between both groups. Tarsiers are found in Indonesia and the Philippines, and they are named for their special elongated tarsal bones, which form their ankles and enable them to leap up to three metres from tree to tree. Nocturnal hunters, their large, mobile eyes are their most notable feature, each one being bigger than the tarsier's brain.

Prosimians

Features common to most prosimians:

- Tooth comb and grooming claw
- Wet naked nose with whiskers
- Arboreal (tree dwelling)
- Grasping hands and feet
- Long, mobile limbs
- Quadrupedal
- Binocular vision
- Upright sitting position
- Nails instead of claws on most digits
- Simpler placenta

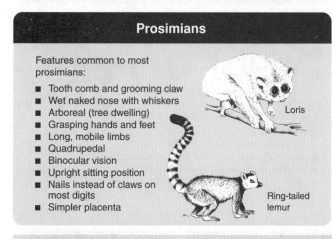

Loris

Ring-tailed lemur

Tarsiers

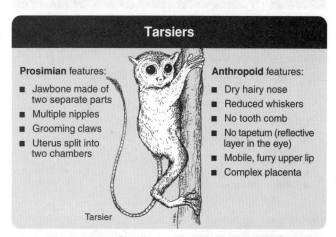

Prosimian features:

- Jawbone made of two separate parts
- Multiple nipples
- Grooming claws
- Uterus split into two chambers

Anthropoid features:

- Dry hairy nose
- Reduced whiskers
- No tooth comb
- No tapetum (reflective layer in the eye)
- Mobile, furry upper lip
- Complex placenta

Tarsier

The Primates

Ring-tailed lemurs are more terrestrial than other lemurs and live in groups.

Slow lorises prefer forest edges, which have plentiful supports and insect prey.

David Haring/Duke Lemur Center

The nocturnal aye-aye is the most structurally specialised of the lemurs.

Tarsiers hunt animal prey at night. They move by clinging and leaping.

1. Describe the significance of the following three features of prosimians that are absent in anthropoid primates:

   (a) Moist rhinarium (structure of the nasal area): _____

   _____

   (b) Reflective tapetum (structure of the eye): _____

   _____

   (c) Dental tooth comb and grooming claw (usually on the second toe of their feet): _____

   _____

2. Describe a feature of tarsiers that is:

   (a) Shared with prosimians: _____

   (b) Shared with anthropoids: _____

   (c) Unique to tarsiers: _____

3. Suggest why biologists have assigned the tarsiers to a taxonomic group separate from the prosimians: _____

   _____

New World and Old World Monkeys

The suborder **Anthropoidea** includes all of the simian primates, i.e. the true monkeys and apes. Unlike prosimians, anthropoids do not have a grooming claw, tooth comb, or a tapetum layer in their eyes. The anthropoids comprise two infraorders: **Platyrrhini** (the New World monkeys) and **Catarrhini** (the Old World monkeys, apes, and hominins). The **New World monkeys** are found in Central and South America and, in general, have rounded nostrils

that face towards their ears (Platyrrhini means flat nose). The **Old World monkeys** of Africa and Asia are diurnal and generally larger than the Platyrrhini primates. Like all catarrhines, they have narrow nostrils that point downwards and flattened nails on their digits. They are divided into two subfamilies: the cheek pouch monkeys, which includes macaques and baboons, and the leaf eating monkeys, which includes langurs and colobus monkeys.

New World monkeys

Superfamily: Ceroidea

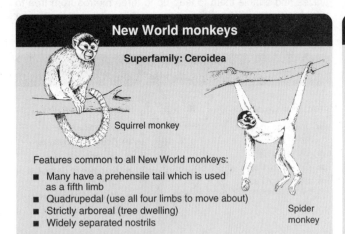

Squirrel monkey

Spider monkey

Features common to all New World monkeys:
- Many have a prehensile tail which is used as a fifth limb
- Quadrupedal (use all four limbs to move about)
- Strictly arboreal (tree dwelling)
- Widely separated nostrils

Old World monkeys

Superfamily: Cercopithecoidea

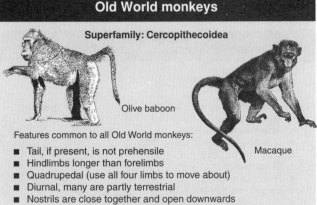

Olive baboon

Macaque

Features common to all Old World monkeys:
- Tail, if present, is not prehensile
- Hindlimbs longer than forelimbs
- Quadrupedal (use all four limbs to move about)
- Diurnal, many are partly terrestrial
- Nostrils are close together and open downwards

Spider monkeys have long, slender limbs and a prehensile tail.

The marmosets (above) and tamarins are all small. The tail is not prehensile.

Macaques are widespread and found in many habitats and climates.

Proboscis monkeys are diurnal and aboreal. Males have a pendulous nose.

1. Describe three features that distinguish Old World monkeys from New World monkeys:

   (a) _____

   (b) _____

   (c) _____

2. Describe one feature of anthropoids that distinguishes them clearly from the prosimians: _____

   _____

3. Describe the geographical distribution of:

   (a) The Old World monkeys: _____

   (b) The New World monkeys: _____

4. Distinguish between the two catarrhine superfamilies, the Old World monkeys and the Hominoidea (apes and hominins):

   _____

   _____

   _____

5. Identify a feature common to all catarrhines: _____

Code: RA 2

Apes and Hominins

Humans are members of the superfamily **Hominoidea**, which includes the apes and hominins (humans and their ancestors). All apes are relatively large primates, with bony eye ridges and flattened noses. The latest taxonomic revisions distinguish the **hylobatids** (the gibbons and siamang) from the **hominids** (humans and the great apes). This new use of the term hominid is explained on page 15, in the activity 'Primate Classification'. The hominids (family Hominidae) are further divided into the subfamilies Ponginae (orangutans) and Homininae (gorillas, chimpanzees, and humans). The hominids have larger bodies and bigger brains than other primates and, unlike the hylobatids or lesser apes, they have no ischial callosities. Hominid males are much larger than the females and are also less aboreal and more terrestrial than the hylobatids, while the orangutan is more aboreal and more solitary than other hominids. Gorillas and chimpanzees walk quadrupedally, using their knuckles to carry the weight of their head and torso. Humans are similar enough to be placed in the same subfamily. Our ancestors diverged from the great apes over 5 mya in Africa by developing bipedal locomotion and then an enlarged brain.

Superfamily Hominoidea

Features of all hominoids (apes & hominins)

- No tail
- Semi-erect or fully erect posture
- Broad chest, pelvis, and shoulders
- Relatively long arms and mobile shoulder joints
- Larger brain

Family Hylobatidae

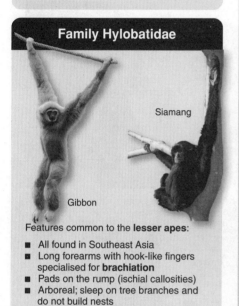

Siamang

Gibbon

Features common to the **lesser apes**:

- All found in Southeast Asia
- Long forearms with hook-like fingers specialised for **brachiation**
- Pads on the rump (ischial callosities)
- Arboreal; sleep on tree branches and do not build nests

Family Hominidae (Hominids)

Subfamily Ponginae

Orangutan

Subfamily Homininae

Gorillini gorillas

All hominini (hominins) have the first three features. The rest are possessed to varying degrees:

- Bipedal: modified feet, thigh, pelvis, and spine
- Large cerebral cortex
- Reduced canines
- Prominent nose and chin
- Reduced brow ridges
- Highly sensitive skin (assumed)
- Body hair very reduced to assist cooling (assumed)
- Complex social behaviour

Australopithecus afarensis

Hominini
humans & their ancestors

Panini
chimpanzees

1. Describe three ways in which apes differ from monkeys:

   (a) _____

   (b) _____

   (c) _____

2. Describe three ways in which the hylobatids differ from the hominids:

   (a) _____

   (b) _____

   (c) _____

3. Discuss how hominins can be distinguished from other hominoids: _____

   _____

   _____

   _____

   _____

   _____

Identifying Primates

The twelve photographs below show representatives from the major primate groups. Use the list provided to assign each one to one of the five **primate groups** listed (write down the letter and name), and **describe three main characteristics** of each group. **Examples**: cottontop tamarin, hamadryas baboon, ruffed lemur, galago or lesser bush baby (appears tarsier-like, but not closely related), lowland gorilla, slow loris, common chimpanzee, gibbon, spider monkey, orangutan, australopithecine, macaque.

1. (a) Identify three **prosimian** examples: _____

   (b) Characteristics: _____

2. (a) Identify two **New World monkey** examples: _____

   (b) Characteristics: _____

3. (a) Identify two **Old World monkey** examples: _____

   (b) Characteristics: _____

4. (a) Identify four **ape** examples: _____

   (b) Characteristics: _____

5. (a) Identify one **hominin** example: _____

   (b) Characteristics: _____

   _____

General Primate Characteristics

The primates have a combination of features that are unique to their group. All primates have retained five digits in the hands and feet (pentadactyly) although some have one digit markedly reduced (e.g. thumb in spider monkeys). **Nails** are found on at least some digits in all modern primates. Climbing is achieved by grasping (not by using claws) and is aided by tactile pads at the end of the digits. Primates have flexible hands and feet with a good deal of **prehensility** (grasping ability). They have a tendency toward **erectness**, particularly in the upper body. This tendency is associated with sitting, standing, leaping, and (in some) walking. The **collarbone** (clavicle) has been retained, allowing more flexibility of the shoulder joint (the clavicle has been lost in many other quadrupedal mammals as an adaptation to striding). Primates have a generalised **dental pattern**, particularly in the back teeth (molars). Unspecialised teeth enabled primates to adopt a flexible omnivorous diet. The snout is reduced along with the olfactory regions of the brain. Baboons go against this

trend, with a secondary increase in muzzle length. There is an emphasis on **vision**, with visual areas of the brain enhanced, and well developed binocular, stereoscopic vision to provide overlapping visual fields and good depth perception. Colour vision is probably present in all primates, except specialised nocturnal forms. The **brain** is large and generally more complex than in other mammals. Foetal nourishment is more efficient and **gestation** is longer than in most other mammals. Single births are the norm. Infancy is prolonged with longer periods of infant dependency and a large **parental investment** in each offspring. Life span is generally longer than most other mammals and there is a greater dependency on highly flexible **learned** behaviour. Unusually for mammals, adult males of many primate species often associate permanently with the group. On the diagram of the capuchin below, briefly describe the general physical characteristics of all primates as indicated:

The primate pictured is a **white-fronted capuchin monkey** (*Cebus albifrons*) from northern South America. These monkeys inhabit the mid-canopy deciduous, gallery forests.

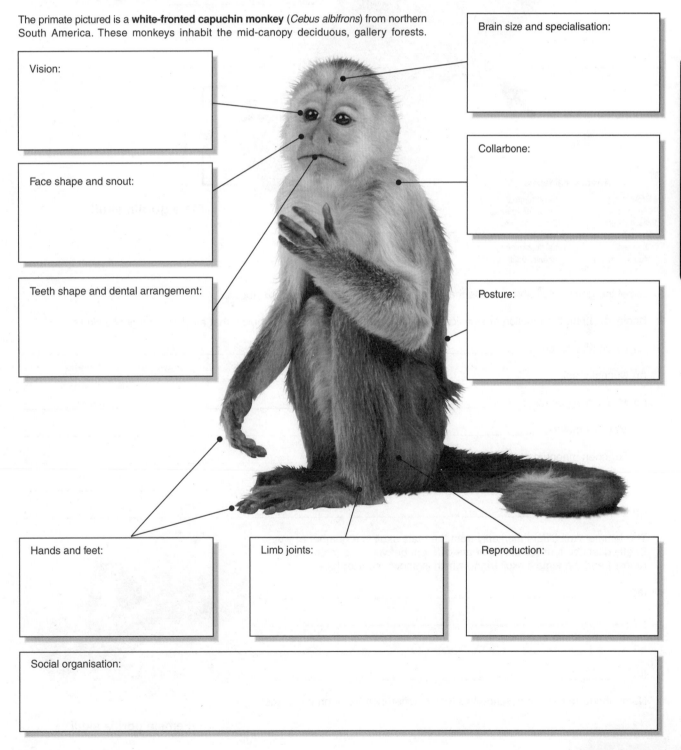

Brain size and specialisation:

Vision:

Collarbone:

Face shape and snout:

Teeth shape and dental arrangement:

Posture:

Hands and feet:

Limb joints:

Reproduction:

Social organisation:

The Primates

Code: A 2

Primate Skull Features

The skull of the gorilla is not typical of primate skulls in general. It is highly specialised for its niche; processing a low-grade diet of foliage. It has been used here simply to acquaint you with the terms used in describing primate skull structures. The skulls on the right are of two apes (a male gorilla and an orangutan), together with that of an Old World monkey (a baboon). The baboon skull is clearly quite different from those of the apes.

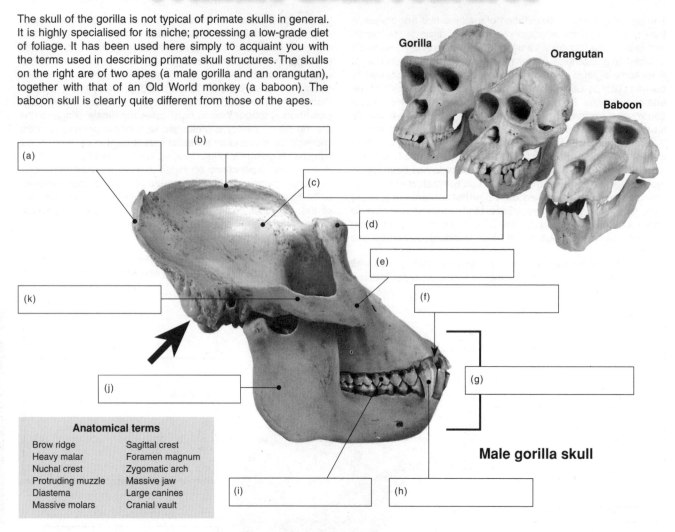

Gorilla

Orangutan

Baboon

(a)

(b)

(c)

(d)

(e)

(k)

(f)

(j)

(g)

(i)

(h)

Male gorilla skull

Anatomical terms

Brow ridge	Sagittal crest
Heavy malar	Foramen magnum
Nuchal crest	Zygomatic arch
Protruding muzzle	Massive jaw
Diastema	Large canines
Massive molars	Cranial vault

1. Label the gorilla skull above, using the list of **anatomical terms** in the box above.

2. Briefly describe the function or significance of each of the **seven** features below that are found on gorilla skulls:

(a) Large sagittal crest: _____

(b) Nuchal crest: _____

(c) Massive zygomatic arch: _____

(d) Massive molars: _____

(e) Foramen magnum at rear: _____

(f) Large canines: _____

(g) Heavy brow ridge: _____

3. The female skull differs markedly from the male skull in a number of ways. Briefly describe four features that are different between the male skull (above) and the female skull (right) which is shown smaller here:

(a) _____

(b) _____

(c) _____

(d) _____

4. Name the term used to describe this type of difference between the sexes:

_____

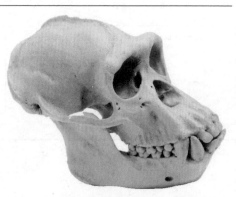

Female gorilla skull

Primate Classification

There is much debate over how the classification of primates should be organised. There are various schemes, some more radical than others, proposed that attempt to interpret the probable evolutionary relationships of the primates. There is considerable evidence suggesting that modern tarsiers are more closely related to simians (monkeys, apes, and humans) than to other prosimians. Many authorities favour an alternative division between the suborder **Strepsirhini** (lemurs and lorises) and the suborder **Haplorhini** (tarsiers and simians). The classification of the apes (hominoids) has its fair share of controversy. There is confusion over the distinction between the families Pongidae and Hominidae within the Hominoidea. Originally, all great apes were placed in the family Pongidae, and the family Hominidae was reserved for humans and their direct fossil relatives. It is now widely acknowledged that the African apes (gorillas and chimpanzees) are more closely related to humans than they are to the orangutans (from Asia), and there have been a number of attempts to reflect this in new classifications. A **revised classification** of the superfamily Hominoidea (opposite and the inset on the following page) places the orangutans in the subfamily Ponginae and combines the African apes and humans in the subfamily Homininae. The outcome of this scheme is that the term 'hominid' would mean 'humans and apes' while the term 'hominin' would mean humans and their ancestors. Note: This revised scheme is the one referred to throughout this workbook.

The table below illustrates the **traditional classification** of primates. The names that apply to each level for our own species (*Homo sapiens*) are shown in bold. Extinct primate groups are shaded solid grey.

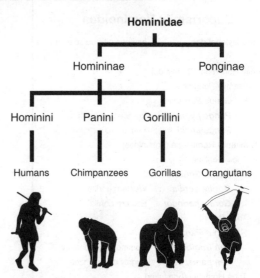

The classification of the great apes and humans is difficult. In the light of DNA comparisons, one solution is to place orangutans in the subfamily Ponginae and to combine African apes and humans in the subfamily **Homininae**. In addition, a new level of classification is created, called *tribes* with the **Hominini** for humans and pre-humans, **Gorillini** for the gorillas and **Panini** for the chimpanzees.

The Primates

Order	Suborder	Infraorder	Superfamily	Family	Examples
Primates	Prosimii (prosimians)	Plesiadapiformes (archaic primates)			extinct plesiadapiformes (*Purgatorius*)
		Lemuriformes	Lemuroidea	Lemuridae Indriidae	lemur indri
			Adapoidea	Adapidae	extinct adapiformes
		Lorisiformes	Lorisoidea	Lorisidae	loris, galago (bush baby)
		Tarsiiformes	Tarsioidea	Tarsiidae	tarsier
				Omomyidae	extinct omomyiformes
	Anthropoidea (simians or anthropoids)	Platyrrhini (New World simians)	Ceboidea (New World monkeys)	Callitrichidae	marmoset, tamarin
				Cebidae (true monkeys)	cebus monkey, spider monkey, howler monkey, capuchins, owl monkey, sakis
		Catarrhini (Old World simians)	Cercopithecoidea (Old World monkeys)	Cercopithecidae	colobus, langurs, macaque, baboon
			Hominoidea (apes and humans)	Oreopithecidae	extinct *Oreopithecus*
				Hylobaitidae	gibbon, siamang
				Pongidae	orangutan, gorilla, chimpanzee
				Hominidae	*Homo sapiens*, extinct australopithecines

1. Provide the classification for the following primates, using the chart above as a guide:

Primate	Order	Suborder	Infraorder	Superfamily	Family
(a) Lemur					
(b) Spider monkey					
(c) Baboon					
(d) Gibbon					
(e) Humans					

Code: DA 3

Superfamily Hominoidea

Family Hylobatidae: {various gibbon species}

Family Hominidae (hominids):
  Subfamily Ponginae
    Genus *Pongo*:
      Pongo pygmaeus   Bornean orangutan
      Pongo abelii   Sumatran orangutan
  Subfamily Homininae (hominines)
    Tribe Gorillini
      Genus *Gorilla*:
        Gorilla gorilla   Western gorilla
        Gorilla beringei Eastern gorilla
    Tribe Panini
      Genus *Pan*:
        Pan troglodytes   Common chimpanzee
        Pan paniscus   Pygmy chimpanzee
    Tribe Hominini (hominins)
      Genus *Homo*:
        Homo sapiens   Human

Genetic relatedness of primates

The diagram below illustrates a new way of classifying the hominoids (apes and humans) based on genetic similarities. The percentages next to each of the points where a split occurs indicates the amount of difference in the total genetic makeup (genomes) of the two groups being considered. For example, the genome of the gibbons compared to the rest of the apes (orangutans, gorillas, chimpanzees) and humans differs by 5.7%. The resulting diagram based on successive splitting of divergent groups is calibrated according to these genetic differences. A large genetic difference between any two groups implies that they are distantly related, whereas small genetic differences suggest they share a recent common ancestor.

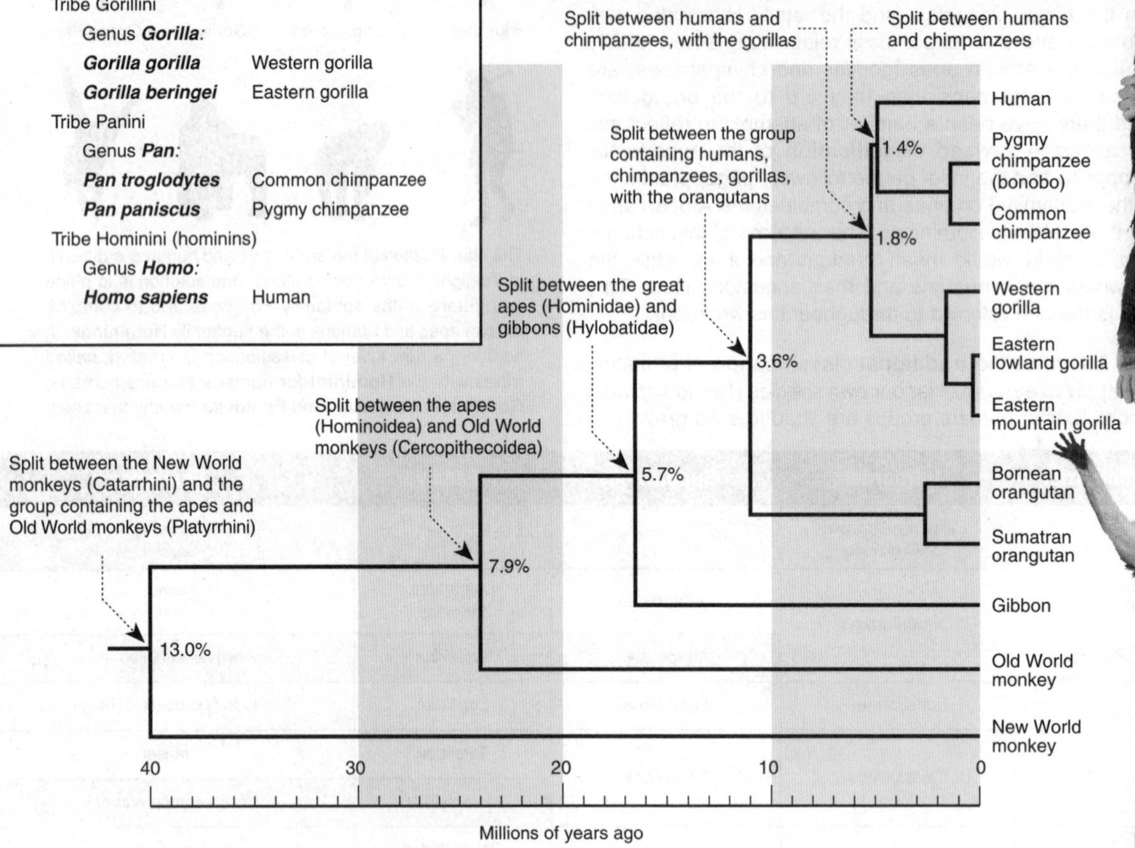

Split between humans and chimpanzees, with the gorillas

Split between humans and chimpanzees

Split between the group containing humans, chimpanzees, gorillas, with the orangutans

Split between the great apes (Hominidae) and gibbons (Hylobatidae)

Split between the apes (Hominoidea) and Old World monkeys (Cercopithecoidea)

Split between the New World monkeys (Catarrhini) and the group containing the apes and Old World monkeys (Platyrrhini)

1.4%

1.8%

3.6%

5.7%

7.9%

13.0%

Human
Pygmy chimpanzee (bonobo)
Common chimpanzee
Western gorilla
Eastern lowland gorilla
Eastern mountain gorilla
Bornean orangutan
Sumatran orangutan
Gibbon
Old World monkey
New World monkey

40   30   20   10   0

Millions of years ago

1. According to the diagram above, showing relatedness according to genetic similarity:

    (a) Identify the hominoid group that is most closely related to the two chimpanzee species: _____

    (b) Name the two chimpanzee species: _____

2. Determine from the diagram how long ago:

    (a) The two species of chimpanzee split from a common ancestor: _____

    (b) The chimpanzees split from the line to humans: _____

    (c) The African apes (and humans) split from the Asian apes (orangutans and gibbons): _____

3. Explain what assumption must be made in order for the degree of genetic diversity to be used as a measure of evolutionary distance:

    _____

    _____

4. Describe two major departures that emerge from this new classification system over the older, traditional one:

    (a) _____

    (b) _____

Physical Features of Primates

For this exercise, view video material on primates, or visit a zoo with primate species on display. Visits are best made early in the morning when these animals are most active. If access to a zoo or video material is not possible, primate physical features could be studied by reference to a good source book (see reference list). The Order Primates includes about 200 species and is unusual among mammals in that living representatives range from primitive forms (e.g. lemurs) that are similar to some of the earliest fossil primates, to advanced forms (e.g. apes) that are probably similar to the immediate human ancestors. By comparing the physical structure of different primates we can arrive at an evolutionary sequence that ends in humans. This does not mean that humans evolved from these modern primates, but from ancestors that probably resembled them.

Distinguishing Physical Features of Primates

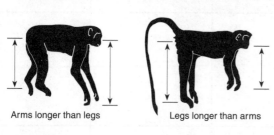

Arms longer than legs — Legs longer than arms

Limb length: Relative length of forelimbs and hindlimbs.

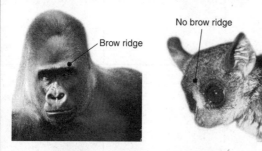

Brow ridge — No brow ridge

Brow ridges: Relative prominence of the ridge of bone above the eyes.

Zone of overlap

Poor binocular vision — Good binocular vision

Eyes: Directed forwards or sideways.

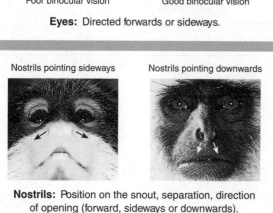

Nostrils pointing sideways — Nostrils pointing downwards

Nostrils: Position on the snout, separation, direction of opening (forward, sideways or downwards).

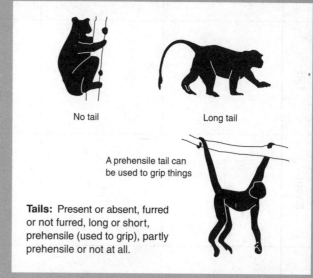

No tail — Long tail

A prehensile tail can be used to grip things

Tails: Present or absent, furred or not furred, long or short, prehensile (used to grip), partly prehensile or not at all.

Long snout — Rounded snout

Split upper lip Whiskers — Continuous upper lip

Snout and lips: Snout obvious or reduced, pointed, rounded or flattened. Whiskers present or absent. Upper lip divided or continuous.

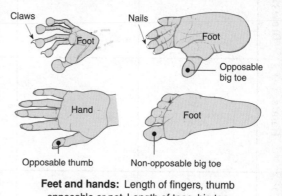

Claws — Foot — Nails — Foot — Opposable big toe

Hand — Foot

Opposable thumb — Non-opposable big toe

Feet and hands: Length of fingers, thumb opposable or not. Length of toes, big toe opposable or not, nails on all hands and feet.

The Primates

1. Study the features above to identify the distinguishing physical characteristics of each of the major groups of primate.
2. Fill in the record sheet on the following page with reference to these guidelines.

Code: PA 3

Primate Physical Features – Record Sheet

Name of primate →	Prosimian	New World monkey	Old World monkey	Ape	Human
Limbs					
Tail					
Snout					
Nostrils					
Eyes					
Brow ridges					
Hands					
Feet					

Primate Niches

While most primates live in tropical rainforests, some species live in sclerophyll forest, mangroves, coniferous forests, or non-forested areas. Baboons live in the savannah and woodlands of Africa, langurs in the dry thorn-scrub of India, and Japanese macaques in the mountains of northern Honshu. Primate sizes range from the 10 cm/55 g red mouse lemur to the 1.5 m/175 kg male gorilla. Primates have diverse ways of moving about. Each enables exploitation, through physical adaptation, of a particular aspect of the environment. Primates also show dietary flexibility. Essentially, they are fruit eaters, but smaller primates supplement fruit with animal matter, and the larger species usually have some leaf-eating adaptations. To illustrate their niche differentiation, describe the mode of locomotion and diet of the primate groups illustrated. Choose answers from the lists provided below:

Modes of locomotion:

Primate locomotion can be divided into six general categories: **arboreal quadrupedalism** (walking and running on all fours along branches); **terrestrial quadrupedalism** (moving on all fours on the ground); **knuckle-walking** (a variation of previous one); **leaping** (moving between tree trunks and branches by rapid extension of the legs); **suspension** (hanging below branches, including brachiation); **bipedalism** (walking and running on two limbs).

Types of diet:

Omnivore, faunivorous (eats insects or mammals), seed eater, flower eater, frugivore (fruit eater), gum and sap eater, foliovore (leaf eater).

Locomotion:

_____

_____

Diet: _____

_____

_____

(e) Gibbons

Locomotion:

_____

_____

Diet: _____

_____

(a) Lemurs and lorises

Locomotion:

_____

_____

Diet: _____

_____

_____

(f) Orangutans

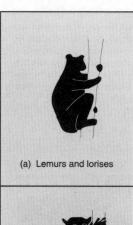

Locomotion:

_____

_____

Diet: _____

_____

_____

(b) Tarsiers

Locomotion:

_____

_____

Diet: _____

_____

_____

(g) Gorillas

Locomotion:

_____

_____

Diet: _____

_____

_____

(c) New World monkeys

Locomotion:

_____

_____

Diet: _____

_____

_____

(h) Chimpanzees

Locomotion:

_____

_____

Diet: _____

_____

(d) Old World monkeys

Locomotion:

_____

_____

_____

Diet: _____

_____

_____

(i) Humans

Code: R 1

Primate Behaviour

Prolonged infant dependency

Primate infants (e.g. the infant **olive baboon** with its mother above) have a long period of parental dependency. This provides food and protection, and offers more opportunity for the young to learn from their parents. In apes and humans, the period of dependency is longer and allows the development of culture.

Grooming behaviour

Grooming of other group members is a social behaviour that has little to do with hygiene. It involves the careful removal of flakes of skin and grass seeds. It helps to create a bond between individuals in a family group and is thought to reduce stress. Here two **crab-eater macaques** from Asia are seen grooming.

Visual and olfactory signals

Ring-tailed lemurs are prosimians from the island of Madagascar, off the east coast of Africa. They communicate using a combination of visual signals and scent marks to map out their territory. Their bushy tail with distinctive markings is held erect like a flag for other members of the family group to see.

Types of Primate Locomotion

Monkey locomotion

Monkeys, such as the **olive baboon** above, walk **quadrupedally** on the palms of their hands and the soles of their feet. In the trees, they walk along the tops of branches, gripping them with their hands and feet.

Brachiation in gibbons

Gibbons are the smallest of the apes and are specialised to use **brachiation** (a technique of under-branch swinging), in combination with rapid climbing, midair leaps, and bipedal running, to move quickly through the forest.

Knuckle walking in chimpanzees

Chimpanzees and **gorillas** spend more time out of trees than do either of the Asian apes. The chimpanzee above shows typical **knuckle-walking** behaviour. The relatively long arms facilitate this mode of locomotion.

1. Discuss types of primate locomotion: _____

_____

_____

_____

2. Describe one feature of primate biology that is important in human sociality and give a reason for your answer:

_____

_____

Primate Behaviour Study

In this exercise you will record behavioural patterns of some nonhuman primates and relate these to humans. To compare the behaviour of different groups of primate, you should observe:

1. A prosimian (if available)
2. A New World monkey
3. Two Old World monkeys
4. At least two ape species

Include in the animals studied at least one species that has a fairly large number of individuals in one enclosure. You should spend at least **15 minutes** observing the behaviour of each species. Use the record sheet on the following page to record your observations. The following is a guide to some of the observations you can make. Write your observations in the spaces provided opposite, and note any behaviours that you think may be relevant.

Many primates have a complex social organisation, but a few do not form social groups. Their social behaviour depends on activity. Nocturnal primates (e.g. many prosimians) either live alone or in monogamous family groups. Diurnal species usually live in relatively large, stable groups, ranging in size from family groups to large bands (e.g. baboons). In primates, communication by smell is most prevalent among prosimians, and to a lesser extent the New World monkeys. Old World monkeys, apes, and humans have a remarkably reduced sense of smell and rely more on visual and vocal signals. These can change rapidly, may convey complex information, and are instantaneous. Chemical signals, by contrast, are relatively slow, spread in all directions and persist.

Behaviour Examples

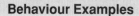

Locomotion	Quadrupedal, bipedal, knuckle-walking, suspension, brachiation (branch-swinging), climbing, branch-leaping, use of prehensile tail
Feeding	Seizing of food by muzzle or hands, breaking of food by hands, muzzle or teeth, storage of food in cheek pouches, movement of jaw, lips, teeth
Vocalising	Vocal or instrumental (such as palm-slapping or chest beating)
Body language	Body posture: submissive, aggressive, 'presenting'. Colour pattern showing sex differences, movements including: swaying, head-bobbing, tail-lashing, eye-contact and smell
Facial displays	Facial expressions; moods expressed such as aggression, submission, anger, threat, pleasure, fear (e.g. baring of teeth, eye-lid flashing)
Social behaviour	Mutual grooming, order of flight, food dominance, play, territorial defence, sexual behaviour
Behaviour towards young	Maternal behaviour, attitude of other group members

The forest-dwelling mandrill baboons exhibit obvious sexual dimorphism. Apart from the difference in size, the male's face is brightly coloured with blue muzzle, red nose, white whiskers and orange beard. In the photo above, a female is presenting herself to the male to initiate mating.

An infant primate depends on its mother for all its food for 2 to 6 months after birth, depending on its size. Most infant primates are carried by their mothers for a further 6 to 12 months. They will depend on them (and perhaps close relatives) for support during fights and protection from danger for 3 to 4 years.

All baboon-like primates have long, sharp canines in long jaws, used in frequent exaggerated yawns. This is part of their social regulatory system.

The "moustache" of the emperor tamarin makes head movements more conspicuous. This makes body language messages less ambiguous.

Most primates do not specialise on one plant but selectively feed on a variety of species, choosing plant parts with low toxicity (e.g. new leaves).

The Primates

Code: PA 3

Primate Behaviour – Record Sheet

Name of primate →	Prosimian	New World monkey	Old World monkey	Ape	Human
Locomotion					
Feeding					
Vocalising					
Body language					
Facial display					
Social					
Young					

Adaptive Radiation in Primates

Recent discoveries of new fossils have clarified our current picture of primate evolution. There are still areas of disagreement; it is uncertain whether the omomyiformes were ancestral to the tarsiers and the anthropoids, or whether they constitute an evolutionary 'dead end' (as portrayed below). The gibbon was the earliest of the ape lines to diverge from the line that lead to humans. There is general agreement from the phylogenetic analyses that *Sivapithecus* is the ancestor of the living orangutan; now the only living representative of the group of Miocene apes called the **ramamorphs**. The **dryomorphs** were probably ancestral to the later hominoid groups, although the position of *Dryopithecus* has long been controversial. The

latest phylogenetic analyses indicate *Dryopithecus* is closely related to *Ouranopithecus*, and that one of these two European genera was the likely ancestor of modern African apes and humans. There is still great debate about the timing of the split between the African apes (chimpanzees and gorillas) and the early humans. Genetic comparisons of the modern apes and humans suggest a recent split. This has been supported by the discovery of *Ardipithecus ramidus* dated at about 4.4 million years. It had strong chimpanzee-like features, but also exhibited the beginnings of some human features as well. The discovery of the 6-7 million year old *Orrorin tugenensis* and *Sahelanthropus tchadensis* possibly represent the earliest hominins.

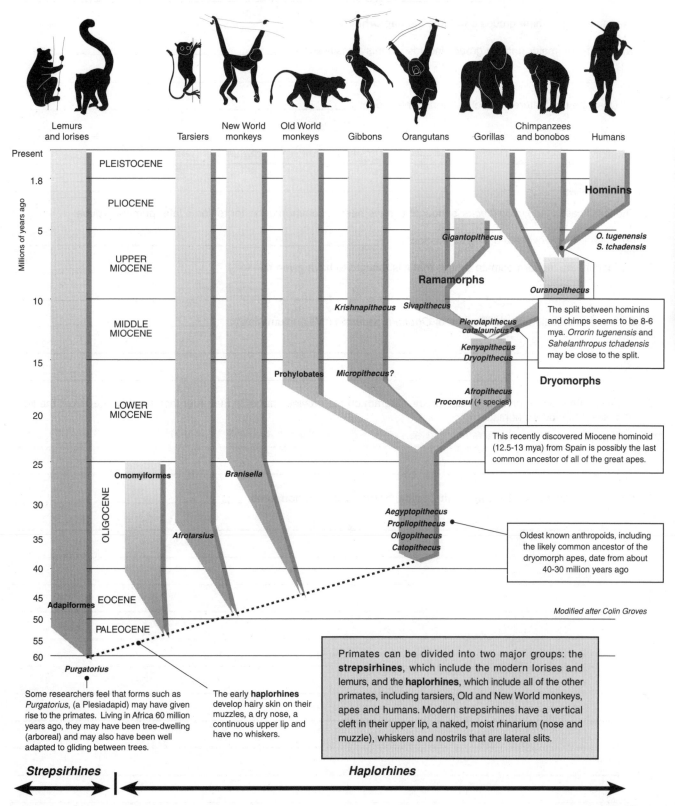

The split between hominins and chimps seems to be 8-6 mya. *Orrorin tugenensis* and *Sahelanthropus tchadensis* may be close to the split.

This recently discovered Miocene hominoid (12.5-13 mya) from Spain is possibly the last common ancestor of all of the great apes.

Oldest known anthropoids, including the likely common ancestor of the dryomorph apes, date from about 40-30 million years ago

Modified after Colin Groves

Some researchers feel that forms such as *Purgatorius*, (a Plesiadapid) may have given rise to the primates. Living in Africa 60 million years ago, they may have been tree-dwelling (arboreal) and may also have been well adapted to gliding between trees.

The early **haplorhines** develop hairy skin on their muzzles, a dry nose, a continuous upper lip and have no whiskers.

Primates can be divided into two major groups: the **strepsirhines**, which include the modern lorises and lemurs, and the **haplorhines**, which include all of the other primates, including tarsiers, Old and New World monkeys, apes and humans. Modern strepsirhines have a vertical cleft in their upper lip, a naked, moist rhinarium (nose and muzzle), whiskers and nostrils that are lateral slits.

Strepsirhines   *Haplorhines*

The Primates

Code: DA 3

1. (a) Identify the earliest known primate fossil: _____

    (b) State the approximate dating for this fossil: _____

2. The **strepsirhines** are a group of 'primitive' primates with features that distinguish them from the other primates.

    (a) Identify two examples of primates from this group: _____

    (b) Describe the features that characterise this group: _____

    _____

    _____

    _____

3. The remaining primate groups constitute the **haplorhines**.

    (a) Identify the major primate groups included in this classification: _____

    _____

    (b) Describe the features that characterise this group: _____

    _____

    _____

    _____

4. The fossil called *Aegyptopithecus* is thought to have been a common ancestor to some later primate groups.

    (a) State the approximate dating for this fossil: _____

    (b) Identify the modern primate groups that it is thought to have given rise to: _____

    _____

5. Explain the evolutionary significance of *Orrorin tugenensis* ("Millennium Man"): _____

    _____

    _____

6. List all of the modern primate groups (in the diagram on the previous page) in the order that they diverged from the line of descent to modern humans:

    _____

    _____

7. Discuss the differences between **anthropoids**, **hominoids**, and **hominins**: _____

    _____

    _____

    _____

    _____

    _____

    _____

    _____

    _____

    _____

    _____

Ancestors of Modern Apes

Early ape ancestors (dryomorphs) probably emerged from the family Pliopithecidae (below top) during a period of great climatic change at the end of the Oligocene and the start of the Miocene (about 25 mya). However, there is still disagreement over the identity and the evolutionary relationships of many fossils. It seems that there was a much greater diversity of hominid groups in the mid-Miocene than was previously recognised. As greater numbers of specimens (and particularly more complete specimens) are recovered, our understanding of the early stages of human evolution should increase correspondingly. Note the *Sivapithecus* (*Ramapithecus*) is firmly off the line to the hominins and is not an ancestor of humans.

Pliopithecids

There are a group of early apes that were present at the end of the Oligocene (about 28 million years ago). One of them, called *Propliopithecus* (also called *Aegyptopithecus*), is thought to be a primitive ancestral hominoid (ape). A modern form that is thought to have diverged from this group is the gibbon.

Includes: *Pliopithecus, Propliopithecus* (also called *Aegyptopithecus*), *Dendropithecus*

Period: 28 million years ago

Brain size: 400 cc

Height: 1.0 m

Distribution: Egypt

Dryomorphs

This widespread extinct group consisted of a large number of species. Dryomorphs were probably the ancestors of all the great apes and humans. The dryopithecines were early apes that probably evolved in Miocene Africa and reached Europe during the shrinkage of the prehistoric Tethys Sea. They may have been the ancestors of the ramamorphs.

Includes: *Dryopithecus, Proconsul, Kenyapithecus, Rangwapithecus, Afropithecus*

Period: 22-9 million years ago

Brain size: 370 cc

Height: 1.0 m

Distribution: Europe, Africa and Asia

Ramamorphs

Asia's sole living great ape is a member of the sub-family Ponginae. This group probably evolved in Africa but spread to Europe and Asia where it continued to persist from about 17-1 million years ago. Until the early 1980s, it was generally accepted that *Ramapithecus* was a direct ancestor to the modern human lineage. Now reclassified as *Sivapithecus*, it is considered a likely ancestor of the modern orangutan (see right). However, recently discovered leg bones from *Sivapithecus* are not similar to modern hominoids (including orangutans), raising doubts about its links even with orangutans. *Gigantopithecus* (another ramamorph) was the largest primate that ever lived - considerably larger than the modern gorilla.

Includes: *Sivapithecus* (once called *Ramapithecus*, now considered to be the same species), *Gigantopithecus*, and *Ouranopithecus* (a recently discovered species).

Period: 17-1 million years ago

Brain size: Within ape range

Height: 1.0-3.0 m

Distribution: Europe, Africa and Asia

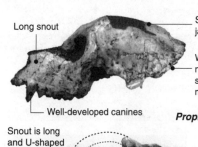

Long snout

Sagittal crest for jaw muscles.

Well-developed nuchal crest for strong neck muscles.

Well-developed canines

Propliopithecus skull

Snout is long and U-shaped

Diastema

Foramen magnum is placed well to the rear of the skull (indicating spine meets the skull at an angle for 4-legged quadrupedal walking).

Teeth are relatively unspecialised with very pointed cusps.

Zygomatic arch is large (pieces of bone broken off in this specimen).

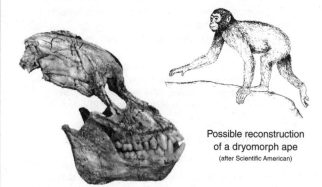

Possible reconstruction of a dryomorph ape
(after Scientific American)

Proconsul skull

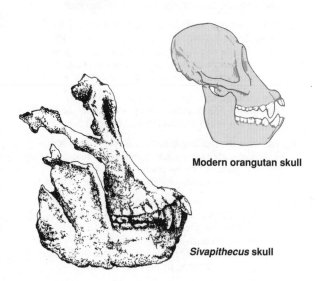

Modern orangutan skull

Sivapithecus skull

ABOVE: A modern orangutan skull has many points of similarity with ramamorph skulls (such as the skull of *Sivapithecus* above)

The Primates

Hominin Evolution

Describing trends in hominin evolution

Evidence for evolution from fossils, DNA, and artifacts; trends exhibited in bipedalism, skull features, and dispersal of <u>Homo</u>.

Learning Objectives

☐ 1. Compile your own glossary from the **KEY WORDS** displayed in **bold type** in the learning objectives below.

Evidence for Hominin Evolution *(pages 28, 55-56)*

☐ 2. Discuss the fossil and other biological evidence contributing to models of hominin evolution and supporting the idea that humans evolved from a bipedal species of African ape. In your discussion:
- Review the way in which fossils are formed, how they are dated, and how geological time is described.
- Describerelative dating techniques using fossil sequence in strata.
- Identify the difficulties associated with interpreting the fossil evidence.
- Explain how evidence from living primates is used to hypothesise about phylogeny. Include reference to: karyotype analyses, DNA sequencing, DNA hybridisation, and use of mitochondrial DNA.

Becoming Human *(pages 28-30, 49-54, 55-56)*

☐ 3. State the full classification of modern humans.

☐ 4. Identify the anatomical and behavioural traits that are unique to humans. Describe the selection pressures on early hominins and the benefits of reducing body hair and adopting **bipedalism** as a form of locomotion.

☐ 5. Describe the anatomical features associated with **bipedalism**. Include reference to the length of the limbs, shape and orientation of the pelvis, **valgus** (carrying) **angle** of the knee, structure of the foot, position of the skull, and curvature of the spine.

☐ 6. In a general way, describe how **brain size** and intelligence has changed during human evolution and explain the selection pressures involved in this.

☐ 7. Describe the regional climate changes in East Africa 7-5 mya, outlining their role in human evolutionary development. Relate the changes of climate to alterations in habitats exploited by primates at the time.

☐ 8. Distinguish between **relative dating** (see #2) and **absolute dating** methods. Identify the different methods by which fossil remains are dated and describe when each of the dating methods is appropriate. Explain how dating methods are used in establishing a timeline for human evolution.

Trends in Hominin Evolution *(pages 31-36, 47-54, 73)*

☐ 9. **Background**: Since the mid-1990s, new fossil finds have overturned earlier ideas about hominin evolution. Altogether the picture is becoming more complicated as new finds uncover more information. Be aware that older textbooks will not reflect recent developments. Note that the species identifications below reflect **currently accepted classifications**. In particular:
- The revised taxonomy of **Hominoidea** (the hominoids) is used throughout this workbook.
- Under this classification scheme, the **Hominidae** (the **hominids**) comprises two Subfamilies, the Pongines (orangutans) and hominines.
- The **Homininae** (the hominines) comprises three tribes: the Gorillini (gorillas), Panini (chimpanzees), and **Hominini** or **hominins** (humans and their bipedal ancestors).
- This classification reflects the closer relationship between humans, chimpanzees, and gorillas.
- Within the tribe Hominini, it is debatable whether *Sahelanthropus tchadensis* is on the line to humans. The debate continues on the exact relationships between the species currently recognised as hominins. Examine alternative views of human evolutionary relationships, with reference to the species listed below (#10, 13, 15) as required.

The first hominins *(also see pages 37-42, 57)*

☐ 10. With the background provided above in mind, describe the anatomical features, geographical distribution, evolutionary relationships, and possible niche differences of the following:
- (a) Earliest hominin genera, e.g. ***Ardipithecus***, ***Orrorin***
- (b) ***Australopithecus*** spp., e.g., Lucy
- (c) The megadonts: ***Paranthropus*** spp.

☐ 11. Explain the terms **robust** and **gracile** in the context of describing early hominin body types. Explain how the evolution of the **australopithecines** was a response to habitat change and a shift in the resources exploited.

Genus *Homo* *(also see pages 43-46, 57)*

☐ 12. Identify the distinguishing characteristics that are unique to the genus *Homo*.

☐ 13. Describe the biological and cultural evolution of the various species belonging to the genus *Homo*:
- (a) *Homo habilis*
- (b) *Homo erectus, Homo ergaster*
- (c) *Homo heidelbergensis* (Archaic *Homo sapiens*)
- (d) *Homo neanderthalensis* (**Neanderthals**)
- (e) *Homo sapiens* (**anatomically modern humans**)

☐ 14. Compare and contrast the hominin taxa listed above (#10, 13), including comparisons of their skeletal structure, cranial capacity, fossil ages and regional locations, and inferred culture.

□ 15. Be aware that a complete listing of hominin species appearing in the scientific literature includes:

- *Sahelanthropus tchadensis* ("Toumai")
- *Orrorin tugenensis* ("Millennium Man")
- **Ardipithecus ramidus** (two subspecies)
- *Australopithecus anamensis, Australopithecus bahrelghazali,* **Australopithecus afarensis, Australopithecus africanus,** *Australopithecus garhi*
- *Kenyanthropus platyops*
- **Paranthropus boisei, Paranthropus robustus**
- **Homo habilis,** *Homo rudolfensis*
- **Homo erectus, Homo floresiensis, Homo ergaster**
- **Homo heidelbergensis** (archaic *Homo sapiens*)
- **Homo neanderthalensis** (Neanderthals)

- *Homo antecessor*
- *Homo sapiens*

□ 16. Suggest why all hominin species, apart from our own, became extinct. Discuss the two main hypotheses for the origin and dispersal of modern humans: **replacement** and **multiregional hypotheses**.

□ 17. In more detail than in #6, identify **specific trends** and consequences of brain development such as continued brain expansion and the development of **Broca's** and **Wernicke's** areas (speech production and recognition).

□ 18. Discuss possible reasons for the loss of body hair, increasing dependence on **group cooperation**, and the development of a language in hominins.

Supplementary Texts

See page 7 for additional details of these texts:

■ Coppens, Y. 2004. **Human Origins: The Story of our Species**, reading as required.

■ Jones *et al.* 1992. **The Cambridge Encyclopedia of Human Evolution**, (CUP), chapters 5-10.

■ Lynch, J. and Barrett, L., 2003. **Walking With Cavemen (BBC)**, as required.

■ Mai, L.I. *et al*, 2005. **The Cambridge Dictionary of Human Biology & Evolution**, (CUP), as required as reference.

Full details are given for the following texts, as they are not listed in the introductory resources section.

■ Burenhult, G., 1993. **The First Humans**, (Uni. of Queensland Press, St. Lucia, QL). *A superb authoritative text covering human origins and history to 10 000 BC.* ISBN: 0-7022-2676-9.

■ Tattersall, I., 1995. **The Last Neanderthal** American Museum of Natural History. *An excellent, thorough resource that extends beyond the scope of the title subject.* ISBN: 0-02-860813-5.

Periodicals

See page 7 for details of publishers of periodicals:

STUDENT'S REFERENCE

■ **Meet Kenya Man** National Geographic, 200(4) Oct. 2001, pp. 84-89. *Description of the recent discovery of Kenyanthropus platyops; a new candidate for humankind's ancestor.*

■ **World of the Little People** National Geographic, 207(4) April 2005, pp. 2-15. *Homo floresiensis, physical and cultural reconstructions.*

■ **Family Ties** National Geographic, 207(4) April 2005, pp. 16-27. *Where the prehistoric find of human remains at Dmanisi fit into human evolution.*

■ **Family Secrets** New Scientist, 19 June 1999, pp. 42-46. *It has long been thought that modern humans wiped out Neanderthals. New evidence suggests this may not have been the case.*

■ **Tracking the First of Our Kind** National Geographic, 192 (3) September 1997, pp. 92-99. *The origins and dispersal of humans. The out of Africa theory is also discussed.*

Presentation MEDIA to support this topic:
HUMAN EVOLUTION

■ **The Greatest Journey** National Geographic, 209 (3) March 2006, pp. 60-73. *Genetic trails left by our ancestors are leading scientists back across time in an epic discovery of human migration.*

■ **The First Americans** New Scientist, 8 April 2006, pp. 42-46. *The controversial discovery of what appears to be human footprints in the desert southeast of Puebla, Mexico, if dated correctly, prove that there were people inhabiting the Americas 30 000 years earlier than first thought.*

■ **Made in Savannahstan** New Scientist, 1 July 2006, pp. 34-39. *In a bold challenge to the conventional story, some argue that hominins migrated out or Africa before H. erectus evolved.*

■ **What Makes Us Different** Time, 9 Oct. 2006, pp. 38-45. *The latest advances in discovering the origin and evolution of humans. Comparing genome maps (analysing DNA samples) of chimpanzees and humans, and soon Neanderthals, can reveal much about each species evolutionary past.*

■ **The Dawn of Humans (series)**
A comprehensive series from **National Geographic** *covering the significant trends and specific stages in the story of human evolution. Titles as follows:*

The Farthest Horizon 188(3) Sept. 1995, pp. 38-51. *Fossils from East Africa are helping paint a portrait of a species on the cusp between apes and humans, Ardipithecus ramidus.*

Face to Face with Lucy's Family 189(3) March 1996, pp. 96-117. *The Australopithecines.*

The First Steps 191(2) Feb. 1997, pp. 72-99. *Trends in human evolution: bipedalism and the selection pressures involved.*

Expanding Worlds 191(5) May 1997, pp. 84-109. *The significance of the Asian fossils and the expansion and evolution of Homo erectus.*

Redrawing Our Family Tree? 194(2) August 1998, pp. 90-99. *In the light of more evidence, the relationships between species in the human evolutionary tree are changing.*

People Like Us 198(1) July 2000, pp. 90-117. *An excellent account documenting human evolution in the last phase of the ice age.*

The First Europeans 192(1) July 1997, pp. 96-113. *An account of human evolution in Europe.*

TEACHER'S REFERENCE

NEW LOOK AT HUMAN EVOLUTION Scientific American **SPECIAL ISSUE** 13(2) July 2003. *A collection of 12 features (some of which are new and some are updated from previous issues):*

■ **An Ancestor to Call Our Own** pp. 4-13. *Controversial new fossils may be close to the origin of humanity (Sahelanthropus, Orririn, Ardipithecus).*

■ **Early Hominid Fossils from Africa** pp. 14-19. *A recently discovered species of Australopithecus anamensis pushes back origins of bipedalism.*

■ **Once We Were Not Alone** pp. 20-27. *For 4 my, many hominid species have shared the planet, often at the same time and in the same region.*

■ **Who Were The Neandertals?** pp. 28-37. *Evidence that these hominids interbred with*

anatomically modern humans and may have had a more advanced culture than previously thought.

■ **Out of Africa Again .... and Again?** pp. 38-45. *Africa is the birth place of humanity. There may have been several waves of hominid emigration out of the continent (an update).*

■ **The Multiregional Evolution of Humans** pp. 46-53. *Both fossil and genetic clues argue that ancient ancestors of various human groups lived where they are found today.*

■ **The Recent African Genesis of Humans** pp. 54-61. *Genetic studies indicate that an African woman of 200 000 ya was our common ancestor.*

■ **Food for Thought** pp. 62-71. *Dietary change was a driving force in human evolution.*

■ **Skin Deep** pp. 72-79. *Human skin colour has developed to be dark enough to prevent UV from destroying folate, while still producing vitamin D.*

■ **The Evolution of Human Birth** pp. 80-85. *The difficulties of childbirth may have created selection pressures for seeking assistance during birth.*

■ **Once Were Cannibals** pp. 86-93. *The practice of cannibalism may be deep-rooted in our history.*

BECOMING HUMAN Scientific American **SPECIAL ISSUE** 16(2) 2006. *A collection of features (previously published) covering aspects of primate behaviour and evolution. Includes:*

Stranger in a New Land pp. 38-47. *New information on the first hominids to leave Africa.*

The Littlest Human pp. 48-57. *The Flores hominid appears to be an island adaptation.*

Founder Mutations pp. 58-67. *Tracing migrations through genetic mutations.*

The Morning of the Modern Mind pp. 74-83. *The origins of human intellect may stretch back further in history than first thought.*

The Emergence of Intelligence pp. 84-92. *Language, foresight and other hallmarks of intelligence are likely to be related to inherent aptitude for certain tasks.*

Internet

See pages 4-5 for details of how to access **Bio Links** from our web site: **www.thebiozone.com**
From Bio Links, access sites under the topics:
HUMAN EVOLUTION: > Human Fossil Record:
• A look at modern human origins • Becoming human • Cavemen: human evolution • Dmanisis site • Early human evolution • Hominid species • Human evolution • Overview of human evolution • Peter Brown's Australian and Asian palaeoanthropology • Virtual skulls at the Australia National Museum • Prominent hominid fossils • Focus on human origins • The human origins program

Hominin Evolution

Human Characteristics

Humans have features that set them apart from other primates. Looking at the differences between modern humans and apes helps to identify a progression of evolutionary changes in the fossils of human ancestors.

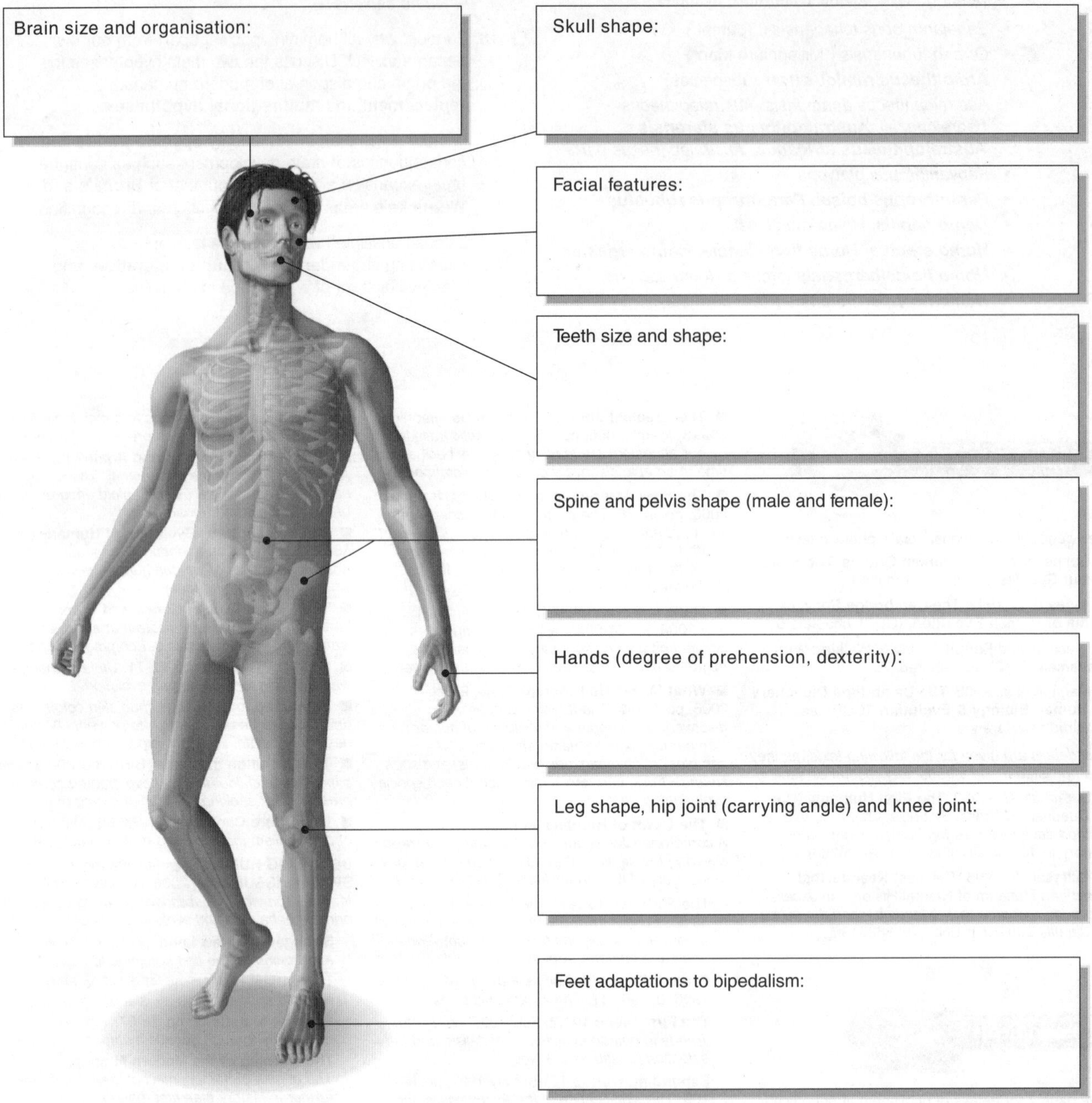

Brain size and organisation:

Skull shape:

Facial features:

Teeth size and shape:

Spine and pelvis shape (male and female):

Hands (degree of prehension, dexterity):

Leg shape, hip joint (carrying angle) and knee joint:

Feet adaptations to bipedalism:

1. In the spaces provided above, describe the characteristics that distinguish humans from other primates.

2. Discuss the significance of **culture**, **abstract thought**, and **social organisation** as attributes that define humans:

_____

_____

_____

_____

_____

_____

_____

_____

_____

_____

Human Skull Anatomy

An understanding of simple skull anatomy is useful when studying the skulls of human ancestors. Knowing the names of the major bones, as well as the features associated with a modern human skull, will help you to identify some of the evolutionary 'landmarks' in the development of humans. Use this page to compare with early human skulls later in this topic.

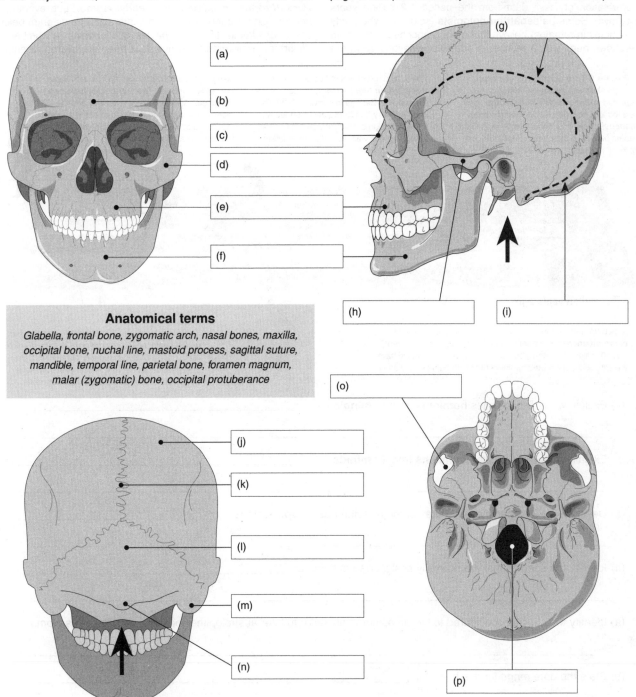

Anatomical terms

Glabella, frontal bone, zygomatic arch, nasal bones, maxilla, occipital bone, nuchal line, mastoid process, sagittal suture, mandible, temporal line, parietal bone, foramen magnum, malar (zygomatic) bone, occipital protuberance

1. Label the three views of a modern human skull using the list of anatomical terms in the box above.

2. Describe six features considered to be characteristic of modern human skulls:

(a) _____

(b) _____

(c) _____

(d) _____

(e) _____

(f) _____

Hominin Evolution

The diagram below shows a provisional 'consensus' view of the family tree for the hominins (the group that includes the modern humans and pre-humans). There is much controversy over the interpretation of fossil data from the period 4-2 million years ago (mya). Some **palaeoanthropologists** (scientists who study fossil hominin remains) believe that more branches existed than are shown here, with a number of adaptive radiations occurring over this period. It is almost certain that the early Australopithecines evolved into *Homo habilis*, which was ancestral to modern humans, by about 2 mya. A divergent branch, genus **Paranthropus**, coexisted with early **Homo**, but eventually became extinct about one million years ago. The diagram below does not attempt to show species relationships. A number of important fossils dated 5-7 mya have been discovered recently.

In 2001, the 6-7 my old remains of a nearly complete skull with gorilla-like features was unearthed in Chad. Nicknamed "Toumai" and assigned to a new genus, **Sahelanthropus tchadensis**, scientists debate whether the skull's features place it in the human family tree, or whether it represents the remains of a proto-gorilla.

The 6 my old remains of five chimpanzee-sized **Orrorin tugenensis** specimens were unearthed at Baringa in Kenya in 2000. The teeth are very humanlike and a perfectly preserved thigh bone clearly shows features associated with walking upright (bipedalism).

Ardipithecus ramidus was an ape with some humanlike features. Two subspecies have been identified: *A. r. ramidus* (4.4 my old) and the older *A. r. kadabba* (5.8 my old). Fossils suggest that it was at least partially bipedal with teeth that were also more humanlike.

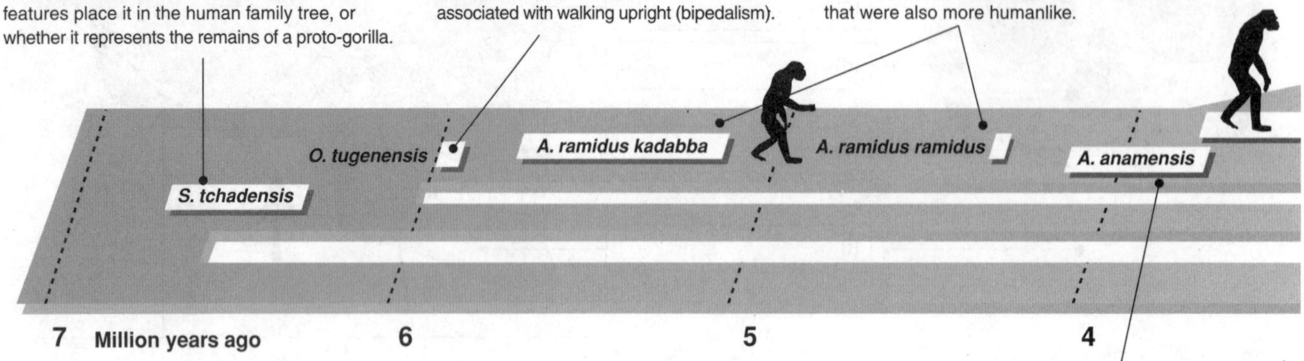

O. tugenensis
A. ramidus kadabba
A. ramidus ramidus
A. anamensis
S. tchadensis

7   Million years ago   6   5   4

New DNA and biochemical evidence suggests that the **last common ancestor** of hominins and apes occurred between 5 and 10 million years ago. The last common ancestor should have a combination of features reminiscent of both humans and apes.

There is a large gap in the fossil record that has until recently been very deficient in early hominin remains. The 1995 discovery of hominin fossils in Kenya, dated about 4 million years old have been named **Australopithecus anamensis**.

1. (a) Explain what distinguishes **hominids** from **hominoids**: _____

_____

(b) Explain what distinguishes **hominins** from **hominids**: _____

_____

2. (a) Describe the key identifying features of the (gracile) **australopithecines**: _____

_____

(b) Identify the species that are normally assigned to this group: _____

_____

3. (a) Identify the species considered to be the common ancestor to later australopithecines and also to genus *Homo*:

_____

(b) State the date range for this hominin: _____

4. People who do not understand hominin evolution often argue that: *"If humans evolved from chimpanzees, then today's chimpanzees should be continuing to evolve into humans everyday"*

(a) State the date range palaeoanthropologists believe hominins and chimpanzees last shared a common ancestor:

_____

(b) Describe two sources of evidence by which researchers have determined this date: _____

_____

(c) Rewrite the statement quoted above to correctly describe the evolutionary relationship between modern chimpanzees and humans:

_____

_____

Unassigned fossil remains dated at 1.8 million years old at **Dmanisi**, in the former Soviet republic of Georgia, suggest an unexpectedly early hominin exodus from Africa.

Homo erectus is a long lived species thought to be the first to venture out of Africa. The recently discovered small hominid *H. floresiensis* appears to have been an offshoot of *H. erectus*, evolving in isolation on the island of Flores in Indonesia.

Modern humans may have appeared more than 160 000 years ago with recently discovered sub-species, *H. sapiens idaltu*. It now seems certain that the **Neanderthals** were not the direct ancestors of modern humans. Recent comparison of DNA suggest that Neanderthals split from the human lineage as long as 600 000 years ago.

Homo habilis species may be too variable to be considered a single species and should be split into two. The additional and older species is **Homo rudolfensis** (1470 skull). Only one of these hominins may have been our direct ancestor.

Australopithecus afarensis is thought to be the common ancestor to both Australopithecines and the genus *Homo*.

H. sapiens

H. hablis

H. ergaster

H. neanderthalensis

H. floresiensis

A. garhi

H. erectus

H. rudolfensis

H. antecessor

A. africanus

H. heidelbergensis

A. bahrelgazali

P. robustus

A. afarensis

P. boisei

K. platyops

Chimpanzees

Gorillas

3 2 1 **Million years ago** 0

Kenyanthropus platyops, a 3.5 my old hominin, coexisted with *A. afarensis*.

Australopithecus bahrelghazali was discovered in Chad in 1995, some 2400 km west of the East African Rift, greatly extending the known geographic range of early hominins.

The **megadonts** (genus *Paranthropus*) are larged toothed vegetarians that disappear from the fossil record at about 1.5 million years ago. The cause of this extinction is not known but it may have been the result of competition with more advanced humans or changes in their habitat.

5. (a) Describe what a **megadont** is: _____

   (b) Identify two species belonging to this group: _____

   (c) Explain why they may have become extinct: _____

   _____

6. Using the diagram above determine:

   (a) The total number of hominin species (i.e. from 6 mya to the present day): _____

   (b) The number of hominin species that existed between 3 and 2 million years ago: _____

7. The recent discovery of the oldest hominin has yielded important information about that stage in human evolution.

   (a) Identify the oldest, clearly described hominin fossil: _____

   (b) Describe two features of this species that suggest that it was not just an ancient chimpanzee: _____

   _____

   _____

8. Explain why it is most unlikely that the Neanderthals were ancestors to modern humans: _____

   _____

   _____

   _____

9. Explain why hominin fossils older than about 4.5 million years have been difficult to find: _____

   _____

   _____

Hominin Evolution

Hominin Skull Identification

Attempt this activity once you have completed the *Hominin Data Sheets*. Give the **name** of the species for each of the hominin skull shapes below. With reference to the following list, describe the **key features** that help in its identification. Features include the overall shape of the skull, the presence of brow ridges, the facial angle, the size (volume) of the brain case compared to the rest of the skull, the presence of a snout, the robustness of the jaw, size of teeth, and the presence of a sagittal crest. Species include: *Homo sapiens, H. neanderthalensis,* archaic *H. sapiens, H. erectus, H. habilis, Paranthropus robustus, P. boisei, A. afarensis, A. africanus.*

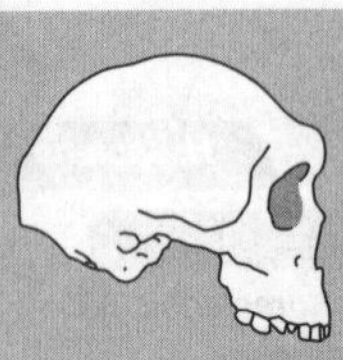

(a) Species: _____

Features: _____

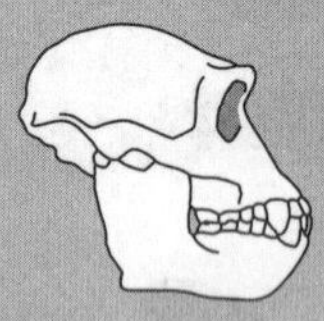

_____

(b) Species: _____

Features: _____

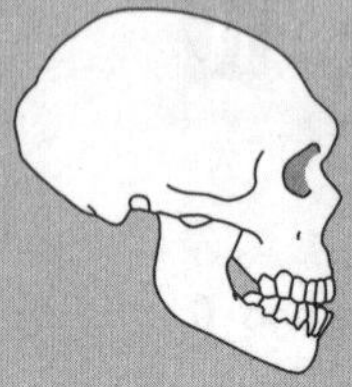

_____

(c) Species: _____

Features: _____

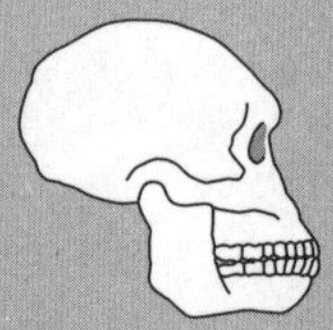

_____

(d) Species: _____

Features: _____

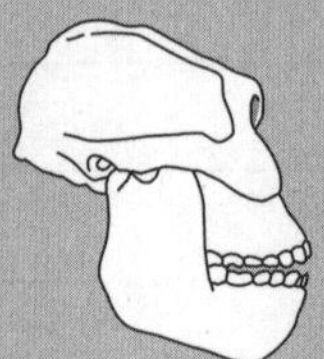

_____

(e) Species: _____

Features: _____

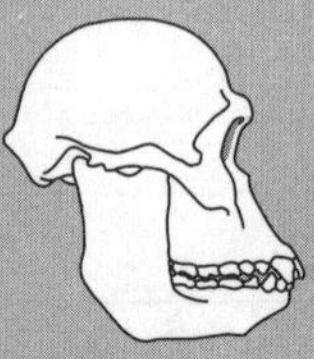

_____

(f) Species: _____

Features: _____

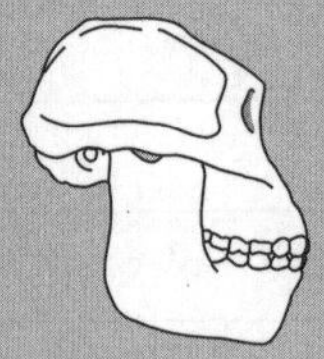

_____

(g) Species: _____

Features: _____

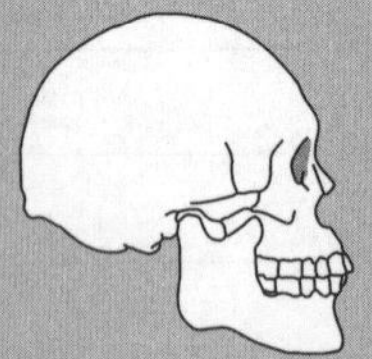

_____

(h) Species: _____

Features: _____

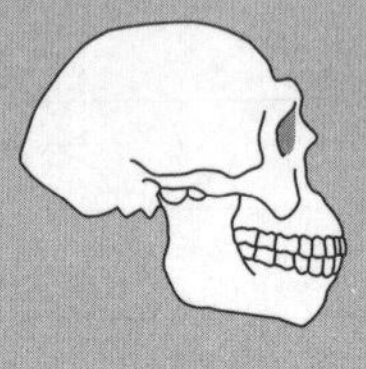

_____

(i) Species: _____

Features: _____

_____

The Emerging View

The view of 'evolutionary tree' illustrated in *Hominin Evolution* is simplified to make it easier to understand where the various hominin groups lie in relationship to each other. There has been a tendency over the last 40 years to try to fit the assembled fossil evidence into a **linear progression** view of human evolution (next page). In the late 1980s and early 1990s, a large number of new hominin fossils were discovered and some of the earlier finds were also reassessed.

This led to an acknowledgement of the more branching nature of the human evolutionary tree. Recently this revised view has been further refined to harmonise the view of human evolution with the evidence gathered on the better understood evolution of other mammals. Human evolution can now be thought of as a succession of **adaptive radiations**, some of which were 'sidelines' to the modern human lineage and all but one species (our own) becoming **extinct**.

Archaic and Modern Humans

Rapid advances in brain size are associated with a suite of new behaviours during this period. Anatomically modern humans emerge from one of the many regional variants and rapidly spread through much of the Old World.

* The species marked with an asterisk (*) were all unknown a decade or so ago (and may be missing from some textbooks on the subject). There are likely to be many as yet 'undiscovered' species in the fossil record between 7 and 4 million years ago.

Paranthropines

These early hominins represent a group specialised for eating a bulky, low-grade vegetarian diet. They developed large cheek teeth, powerful chewing muscles and a generally **robust** skull (large crests for muscle attachment, heavily buttressed face).

Erectines

Big changes in the post-cranial skeleton with body height achieving modern proportions. Marked behavioural changes are matched by increasing brain volume, with sophisticated tool manufacture and use employed to kill and process small sized game. It is still unclear whether *H. floresiensis* belonged to this group.

Habilines

This group shows the first signs of brain enlargement, more meat in the diet as well as the first recognisable stone tool culture. The post-cranial (below the head) skeleton remains small and slight much like that of the australopithecines.

Australopithecines

The earliest of these hominins were among the first apes to achieve bipedalism. They possessed a **gracile** body form and were probably opportunistic omnivores, scavenging meat from carcasses and feeding off a wide range of resources.

Very Early Hominins

Essentially chimpanzee-like animals that have begun to show some human characteristics in their locomotion (bipedalism) and in the shape and arrangement of their teeth. The oldest specimens were only recently discovered; *Orrorin tugenensis* in 2000, *Sahelanthropus tchadensis* in 2001 (although some scientists have suggested that the latter may be a gorilla ancestor).

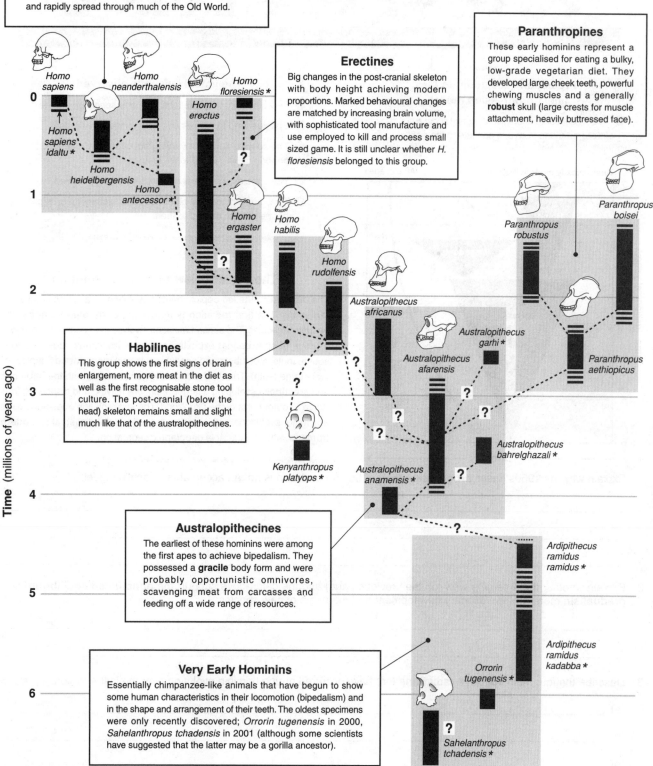

Time (millions of years ago)

Hominin Evolution

Code: RDA 2

A 1960s view of human evolution

The illustration below was in common usage in the popular press 30 years ago to represent the **linear progression** from a primitive apelike ancestor to modern humans. It is still used as a visual metaphor for the idea of evolution in the world of advertising.

Evolving lineage with the accumulation of gradual genetic changes under the influence of natural selection

Predictions according to the linear progression model

● The fossil record should consistently show smooth intergradations from one species to the next.

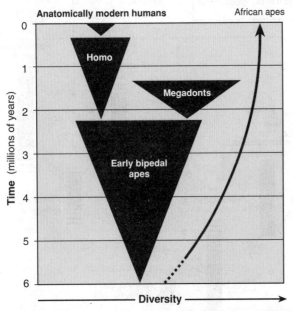

The actual evidence observed

● Few smooth intergradations from one species to the next.

● Species tend to appear suddenly in the fossil record.

● The species linger for varying but often very extended periods of time in the fossil record.

● The species disappear as suddenly as they arrived.

● They are replaced by other species which might or might not be closely related to them.

Source: Robert Foley, (1995) **Humans Before Humanity**, Blackwell Publishers

The current view of human evolution

The diagram on the left depicts human evolution as a series of adaptive radiations. The first radiation is that of the **early bipedal apes**, the australopithecines. The second radiation involves the genus *Paranthropus*, a group of species that exploited a coarse, low-grade vegetable food source (nuts, root tubers and seeds) resulting in **megadontic** adaptations (very large teeth). The third radiation is genus *Homo*, with the **habilines** and **erectines** developing a larger brain, diversifying and dispersing from Africa to other parts of the Old World. The last radiation does not involve any major evolutionary divergence, but reflects the dispersal of **modern humans** with considerable geographic separation.

Source: Ian Tattersall, (1995) **The Fossil Trail**, Oxford University Press

1. Explain why the 1960s **linear progression** view of human evolution is not an acceptable scientific model:

_____

_____

_____

2. Explain whether the *emerging view* (on the previous page) and the data above support the **punctuated equilibrium** or **gradualism** models of evolutionary development and the origin of new species:

_____

_____

_____

3. Describe the four main **adaptive radiations** that have occurred during hominin evolution over the last 4 million years:

(a) _____

(b) _____

(c) _____

(d) _____

Human Evolutionary Models

Various researchers have proposed interpretations of the fossil data to produce a family tree of human evolution, called a **phylogeny**. The interpretation of which fossils are related to others is one of the ultimate goals of palaeoanthropology. A great deal of fossil evidence has been discovered only recently and is still in the early stages of being analysed. This means that their relevance to the overall evolutionary picture is not yet clear, and in some cases their position is no more than a 'best guess' until the analysis is complete. Not only are there fossils still being analysed, but it is inevitable that new finds will be discovered, and these may require slight or significant changes in the evolutionary models proposed. Four of the more recent models that represent current thinking have been portrayed below. There are many reasons why the specialists disagree on the fine detail of hominin evolution. One reason is that it is difficult to recognise which are different fossil species in the fossil record, and which merely represent males and females of the same species. Another reason is differences of opinion as to the meaning or importance of some anatomical features. (Note: the **small** and **large Hadar species** below refer to the species *Australopithecus afarensis* that has been split into two separate species by some proponents).

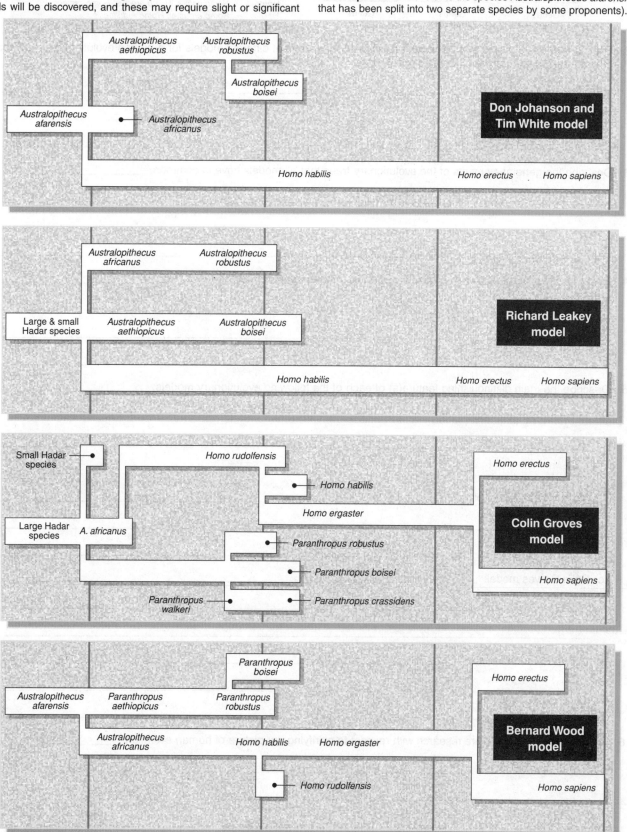

Hominin Evolution

| Years ago: | 3 million | 2 million | 1 million | Present |

Code: DA 3

36

1. Using different coloured highlight pens, colour code the various portions of the evolutionary trees on the previous page using the following key:

    Key to major hominin groups

Genus *Australopithecus*	Genus *Paranthropus*	Genus *Homo*

'Gracile' =	*Australopithecus afarensis, Australopithecus africanus,* Large Hadar species, Small Hadar species
'Robust' =	*Paranthropus robustus (Australopithecus robustus)* *Paranthropus boisei (Australopithecus boisei)* *Paranthropus aethiopicus (Australopithecus aethiopicus)* *Paranthropus crassidens* *Paranthropus walkeri*
Homo =	*Homo habilis, Homo erectus, Homo sapiens, Homo ergaster, Homo rudolfensis*

2. Explain how it is possible that palaeoanthropologists can arrive at different models for a human evolutionary tree when they are taking into account the same fossil data:

    _____

    _____

    _____

3. Describe the general elements of the evolutionary trees that all models have in common: _____

    _____

    _____

    _____

4. Describe the extent to which the models conflict in their interpretation of the fossil data: _____

    _____

    _____

    _____

5. Describe the main distinguishing feature(s) of each of the following evolutionary models:

    (a) Johanson-White model: _____

    _____

    _____

    (b) Richard Leakey model: _____

    _____

    _____

    (c) Colin Groves model: _____

    _____

    _____

    (d) Bernard Wood model: _____

    _____

    _____

6. Discuss the needs of future research with respect to clarifying our picture of human evolution: _____

    _____

    _____

    _____

    _____

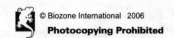

Hominin Data Sheets

This exercise is designed to collate the information on the various hominin species (below) that you may have gathered from a wide range of sources (texts, videos, and class work). The eight data summary sheets can be used to describe key features and points on each species. Descriptions should be brief; in many cases, key words or brief sentences will be all that is required. After completing this activity, attempt *Hominin Skull Identification*. The following species are represented:

Ardipithecus	*Australopithecus*	*Paranthropus*	*Homo*
Ardipithecus ramidus	*Australopithecus anamensis*	*Paranthropus robustus*	*Homo habilis* (including *H. rudolfensis*)
	Australopithecus afarensis	*Paranthropus boisei*	*Homo erectus* (including *H. ergaster*)
	Australopithecus africanus		Archaic *Homo sapiens* (*H. heidelbergensis*)
	Australopithecus bahrelghazali		*Homo antecessor*
	Australopithecus garhi		*Homo neanderthalensis*
			Homo floresiensis
			Homo sapiens

NOTE: No data sheets are provided for recent discoveries including: *Sahelanthropus tchadensis*, *Orrorin tugenensis*, and *Kenyanthropus platyops*. However information on each of these discoveries is provided.

Below are some ideas on how you can analyse the data that you have collected. The data sheets for each hominin have some clearly defined places for answers. Use the 'Additional Notes' box for comments on culture, skeleton, habitat, etc. It is useful to make comparisons between hominins dated immediately before and after the one you are making notes about. Note that extra information is provided here: you are not required to provide detail for every species. The trends are important.

Skull Features

Where indicated, label the main distinguishing features of the skull. Note that skulls are not available for every species. Various features can be considered, but not all will apply to each skull. See the diagram below for help.

1. **Face**
   (a) Size of the face compared to the braincase.
   (b) Degree of prognathism (snout or muzzle development) of the jaw and mid face (mid-face projection).
   (c) Development of brow ridges (supraorbital tori): size, thickness, arching.
   (d) Size of cheek region.

2. **Jaws (mandible)**
   (a) Size and thickness of lower jaw.
   (b) Degree of curvature of dental arcade (tooth row).
   (c) Presence or absence of chin.

3. **Braincase**
   (a) Shape of forehead (slope, height).
   (b) Rear view: where is skull the widest, low down or high up? Shape: pentagonal, rounded, bell-shaped?
   (c) Presence of crests: Nuchal crest for neck muscles, sagittal crest for jaw muscles.
   (d) Shape of occipital region (back of skull) when viewed from the side: presence of bun?
   (e) Dorsal (top) view: where is the skull widest (rear, middle ear, etc.)?
   (f) Position of foramen magnum (opening at base of skull connected to spine).

Diet

Describe the likely food resources utilised by the hominin. This is sometimes conjecture, based on the wear patterns on the surface of the teeth. Other instances provide more direct evidence such as the fossilised or mummified remains of the food. Indicate whether the hominin is completely herbivorous, omnivore, or carnivore. In some cases a species is described as being an 'opportunist', hunter gatherer, or farmer. It has been suggested that the feeding of some early hominins may have involved opting for one of two strategies: a low grade, high volume diet; or a high grade, low volume diet. Diet is probably a major selection pressure that has strongly influenced both physical and cultural evolution.

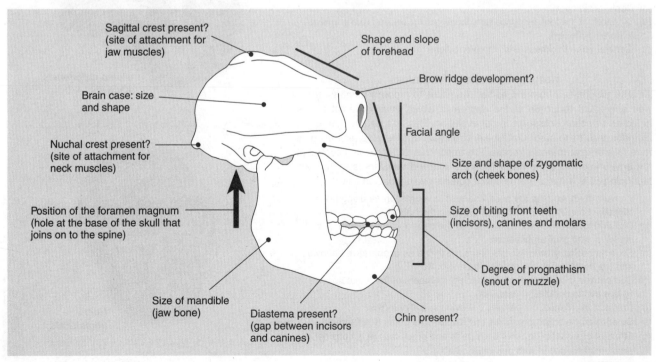

Hominin Evolution

Code: RD 2

Geographic Distribution

List the regions (e.g. east Africa, Asia) or the countries (e.g. Kenya) where fossils of the species have been found to date. The sites are marked on the map with a triangle.

Culture

The various hominin species each have characteristic cultural features. We are restricted to describing those cultural features that have left some record in the form of fossilised remains. So although we find such things as stone tools at a certain age, this does not mean that the early hominins did not use such perishable tools such as wood, bamboo, and other organic materials that failed to preserve. Consider the following points:

(a) Stone tool technology used.
(b) Other materials used (wood, bone, ivory, clay for pottery, copper, bronze, iron, precious metals).
(c) Degree of workmanship required to produce the tool.
(d) Evidence of using fire (e.g. to cook (hearth), to hunt, for security from predators).
(e) Evidence of artistic expression (e.g. rock paintings, carvings, statues) and their significance.
(f) Evidence of abstract thought, spirituality and religion (eg. burials, cannibalism).
(g) Evidence of spoken word (voice box development), written word, higher technologies for communication.

Dentition

The teeth provide important clues about the diet of the hominin. Look for a diastema (gap between the canine and incisors).

(a) Absolute size of teeth (especially molars).
(b) Relative proportions of different tooth types: incisors, canines, molars.
(c) Amount of wear on the teeth; this indicates the quality of the diet: high grade (meat) or low grade (tough plant).

Habitat

Describe the nature of the habitat of the hominin if known. This may be African open savannah for the earlier forms but may include more varied habitats (sub-tropical forests, temperate forests, tundra, and even subarctic) for the later, more widespread hominins.

Skeleton

This section mainly deals with what is called the **post-cranial skeleton**; the skeleton apart from the skull. Various features can be investigated:

(a) Structure of the pelvis: shape, size of birth canal.
(b) Angle of the femur (thigh bone) and the knee joint.
(c) Structure of the spine: curvature and relative size of the vertebrae.
(d) Depth of rib cage.
(e) Structure of the foot; evidence for adaptations for walking and primitive features if present.
(f) General bone thickness and limb proportions.

Natural Selection Pressures

As time progressed from the earliest hominins to modern humans, there was a gradual reduction in the degree to which environmental forces acted as selection pressures in our evolution. This was largely due to our increasing ability to manipulate the environment and develop technologies to reduce environmental stresses. The trend towards increasing control of the environment and the development of learned behaviours to improve technologies is a major theme of our cultural evolution. Consider:

(a) Climatic changes (e.g. ice ages, changes in vegetation resulting from climatic change) and the effect of these on the habitat.
(b) Diet and food resources: their effect on the development of teeth, jaw muscles, and skull architecture.
(c) Food gathering strategies: affecting such things as cooperative behaviour and the development of assisting technologies.
(d) Competition with other hominin species, members of the same species (relate this to population densities).
(e) Predator avoidance: cooperative behaviours, technologies.
(f) Bipedalism: selection pressures that favoured its origin & refinement.
(g) Effect of bipedalism on other body parts and processes (e.g. consider problems associated with our walking habit).

Australopithecus afarensis

Paranthropus aethiopicus
KNM WT 1700

Homo rudolfensis
KNM ER 1470

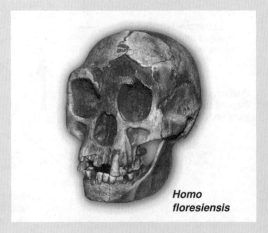

Homo floresiensis

Photos: www.skullsunlimited.com

The Early Hominins of Africa

Sahelanthropus tchadensis

In 2001, the 6-7 my old remains of a nearly complete skull with gorilla-like features was unearthed in Chad. It was nicknamed 'Toumai' and assigned to a new genus, **Sahelanthropus tchadensis**. No bones below the skull have yet been discovered so it is not known whether or not *S. tchadensis* was bipedal. A cranium, mandibular fragment, and four teeth have been recovered from six specimens. Its brain is small, comparable in size to that of a chimpanzee, and the foramen magnum is positioned towards the back of the skull as in apes. The canines are small, and the enamel thickness of the premolars and molars is intermediate between the between those of a chimpanzee and those of a human. Scientists debate whether the skull's features place it in the human family tree, or whether it represents the remains of a 'proto-gorilla' (an early gorilla ancestor).

Orrorin tugenensis

Orrorin tugenensis, or 'Millennium man' was discovered in late 2000. A new hominin from Kapsomin, Kenya, it is claimed as the oldest hominin yet described. *Orrorin* means 'original man' in the local dialect. Fossil evidence for *O.tugenensis* is scant, and only thirteen pieces of bone, consisting of bits of a lower jaw, several teeth, fragments of the arm, thigh bone (femur), and a finger (phalanx), from at least five different individuals have been found. Sediments in which the bones were found have been dated at 6 my old. The size and morphology of the teeth are intermediate between those of a chimpanzee and those of a human. Central incisors are large, the canine is large for a hominin, and the lower fourth premolar is ape-like. Like our molars, the molars of *O.tugenensis* are smaller than of those of any of the australopithicines and have very thick enamel. Grooves in the femora of *O.tugenensis*, presumably where muscles and ligaments attached, suggest the species was bipedal, but a humerus and phalanx suggest aboreal adaptations.

Kenyanthropus platyops

Named in 2001 from a 3.5 mya partial skull found in Lake Turkana, Kenya. This specimen is comparable with *Australopithicus afarensis*, yet quite distinct from it, demonstrating that these two hominins coexisted. A mixture of primitive and advanced traits characterise the two partial skulls attributed to this species. The most distinctive advanced feature of **Kenyanthropus platyops** is the high forward cheekbones, which give the face a flat appearance, similar to that of the later *P. boisei*. Another advanced feature is the teeth, which are smaller than in the australopithecines, but significantly larger than those of *Orrorin*. A primitive trait is a small ear canal similar to *Ardipithecus ramidus* and *Australopithecus anamensis*.

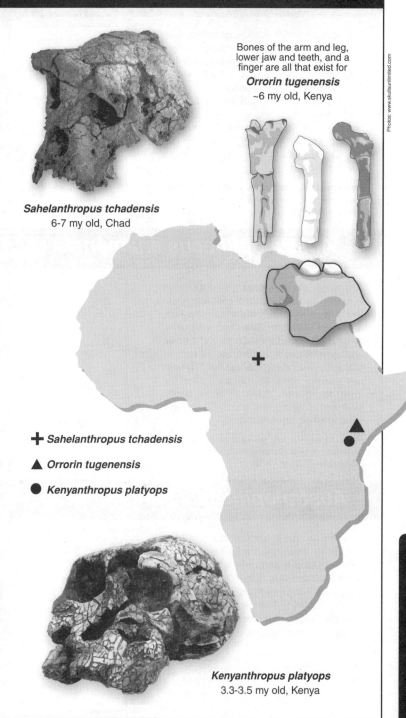

Bones of the arm and leg, lower jaw and teeth, and a finger are all that exist for
Orrorin tugenensis
~6 my old, Kenya

Sahelanthropus tchadensis
6-7 my old, Chad

✛ *Sahelanthropus tchadensis*

▲ *Orrorin tugenensis*

● *Kenyanthropus platyops*

Kenyanthropus platyops
3.3-3.5 my old, Kenya

Human Evolution

1. Using the information given in the data sheets on this page and the following page, complete the timeline of early hominin evolution to show the relative positions of each hominin. *Ardipithecus ramidus* has been done for you.

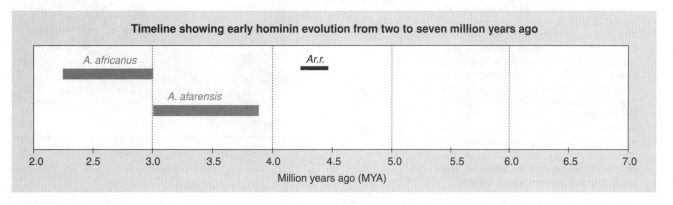

Timeline showing early hominin evolution from two to seven million years ago

Ardipithecus ramidus

This species, which was discovered in 1994, was originally thought to be an early hominin, with limited evidence for bipedalism. It was dated at 4.4 mya and was originally named *Australopithecus ramidus*. Now reclassified under a new genus, it is regarded by some researchers as an ape with some unusual characteristics. Some individuals may have been about 1.2m tall. Other fossils found with this hominin indicate that it may have been a forest dweller.

For an excellent update on the status of this species, see the article: 'Bones of Contention' in New Scientist, 4 November 1995, p. 14-15.

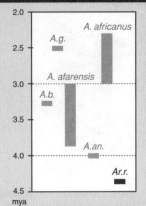

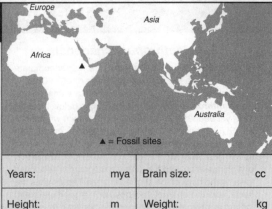

▲ = Fossil sites

Years:	mya	Brain size:	cc
Height:	m	Weight:	kg

Geographic distribution:

Australopithecus anamensis

Anamensis was discovered at Kanapoi, Kenya in 1994. The find consists of complete upper and lower jaws, teeth from several individuals, a piece of skull, arm bones and a leg bone. *Anamensis* existed between 4.2 and 3.9 mya and had a mixture of primitive, ape-like features and advanced, human-like features. The teeth and jaws are similar to older fossil apes. The lower leg bones, however, show strong evidence of bipedalism and the upper arm bone is extremely human-like.

For an excellent article providing background, see the article: 'Bones of Contention' in New Scientist, 4 November 1995, p. 14-15.

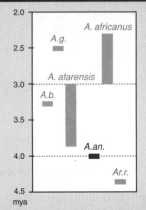

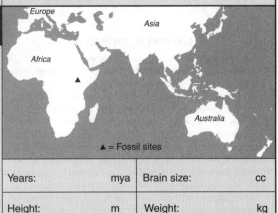

▲ = Fossil sites

Years:	mya	Brain size:	cc
Height:	m	Weight:	kg

Geographic distribution:

Australopithecus bahrelghazali

An important discovery, with characteristics similar to *Australopithicus afarensis*. *A. bahrelghazali* was discovered in Chad in 1995, some 2,400 km west of the East African Rift, greatly extending the known geographic range of early hominins. The fossil find consists of a partial jawbone and teeth dated at about 3.5 to 3.0 million years ago.

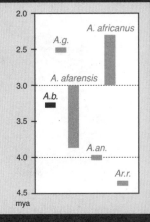

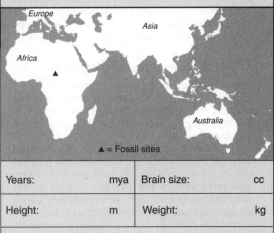

▲ = Fossil sites

Years:	mya	Brain size:	cc
Height:	m	Weight:	kg

Geographic distribution:

Australopithecus garhi

This important find, made in Ethiopia and named in 1999, is known from a partial skull. This skull, dated at about 2.5 mya, differs from other species of *Australopithecus* in its combination of features: the primitive skull shape and extremely large size of the teeth (especially molars). The remains of two other hominins, probably less than 1.5m tall, were found nearby. They are also dated at 2.5 mya and may be from the same species. When arm bones and leg bones are compared, they show a mix of human and ape-like proportions (humerus to femur ratio is human-like, while upper arm to lower arm ratio is ape-like).

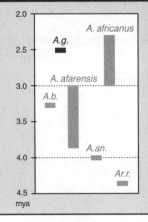

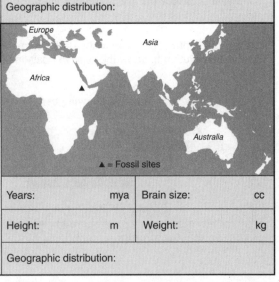

▲ = Fossil sites

Years:	mya	Brain size:	cc
Height:	m	Weight:	kg

Geographic distribution:

Australopithecus afarensis

Small, gracile, small-brained, and bipedal, *Afarensis* existed between 3.9 and 3.0 mya. The skull is similar to that of a chimpanzee, except for more human-like teeth. Brain size varied between 375-550 cc. The humanlike pelvis and leg bones confirm they were bipedal. Height ranges from 1.0 to 1.5 m (sexual dimorphism). Some researchers claim such differences in height suggest two separate species, not sexual dimorphism. Nicknames: Lucy, The First Family, Laetoli footprints.

Artist's reconstruction

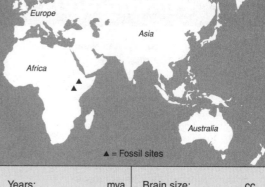

▲ = Fossil sites

Years:	mya	Brain size:	cc
Height:	m	Weight:	kg

Diet:

Geographic distribution:

Additional notes:

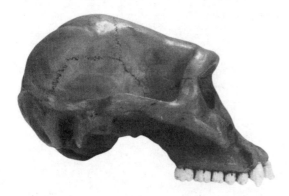

Composite reconstruction

Australopithecus africanus

Africanus existed between 3.0 and 2.0 mya. Similar to *afarensis*, it was also small, gracile, and bipedal, but slightly larger in size. Brain size may also have been slightly larger, ranging from 420 to 500 cc. Generally considered to be specific to South Africa. Differs from the early australopithecines in east Africa by having larger back teeth and smaller canines. The jaw shape is fully human-like. Nicknames: Taung baby, Mrs Ples.

Artist's reconstruction

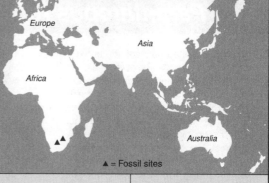

▲ = Fossil sites

Years:	mya	Brain size:	cc
Height:	m	Weight:	kg

Diet:

Geographic distribution:

Additional notes:

Sts 5 skull from Sterkfontein in South Africa

Hominin Evolution

42

Paranthropus boisei

One of a group of robust species of early hominin. Existing between 2.1 and 1.1 mya, it had a brain size ranging from 500 to 545 cc. Known for its very massive jaws, large molars, and attachments on the skull associated with chewing muscle. Probably fed on a tough diet of low grade foods: tubers, grains, and other plant material. This species is also referred to by some researchers as *Australopithecus boisei*. Nicknames: Zinjanthropus, Nutcracker Man.

Artist's reconstruction

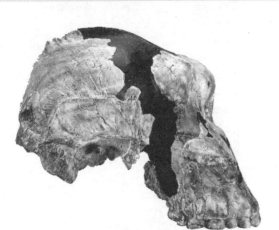

▲ = Fossil sites

Years:	mya	Brain size:	cc
Height:	m	Weight:	kg

Diet:
Geographic distribution:
Additional notes:

OH 5 found at Olduvai Gorge in Tanzania

Paranthropus robustus

Robustus existed between 2.0 and 1.5 mya and had a brain size of about 530 cc. It had a similar body to that of *africanus*, but a larger and more robust skull and teeth. The massive face was flat or dished, with large brow ridges and no forehead. Massive grinding teeth set in a large jaw suggest that it probably fed on a diet of tough, coarse plant food that needed a lot of chewing. May have used bones as digging tools. Some researchers classify this species as *Australopithecus robustus*.

Artist's reconstruction

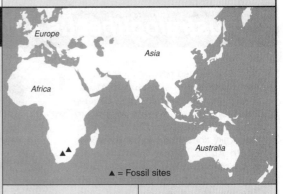

▲ = Fossil sites

Years:	mya	Brain size:	cc
Height:	m	Weight:	kg

Diet:
Geographic distribution:
Additional notes:

SK 48 from Swartkrans in South Africa

Homo habilis

Habilis existed between 2.4 and 1.5 mya. Although similar to australopithecines in many ways (e.g. height of males is 1.3 m), their brain size was considerably larger (500 to 800 cc). Some researchers argue that this species is too variable in its present classification. Instead, they propose that it be split into 2 species: *Homo habilis* (**ER-1813**, shown below) and the more robust *Homo rudolfensis* (**ER-1470**). One *habilis* brain cast shows a bulge of the Broca's area, suggesting rudimentary speech.

Artist's reconstruction

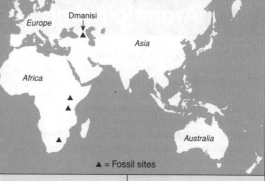

▲ = Fossil sites

Years:	mya	Brain size:	cc
Height:	m	Weight:	kg

Diet:

Geographic distribution:

Additional notes:

The recent find in Dmanisi, Georgia, is the first and only record (to date) of a habilis type outside Africa.

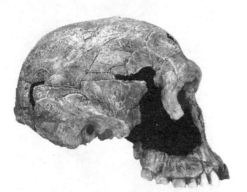

KNM-ER 1813 skull from Koobi Fora region to the east of Lake Turkana, Kenya

Homo ergaster

Larger brained than previous *Homo* species, with volumes of 850 to 1000 cc. Previously considered to be part of *Homo erectus*, but now thought to be a separate species. *Homo ergaster* refers to what used to be called early African forms of *Homo erectus*, existing 1.8 to 1.4 mya. May have ventured out of Africa 1.7 mya (Dmanisi, Georgia). Earliest hominid with human-like body proportions. An nearly complete skeleton of a 9 year old boy was 1.6 m tall (estimated 1.85 m as an adult). Nickname: Turkana Boy.

Artist's reconstruction

▲ = Fossil sites

Years:	mya	Brain size:	cc
Height:	m	Weight:	kg

Diet:

Geographic distribution:

Additional notes:

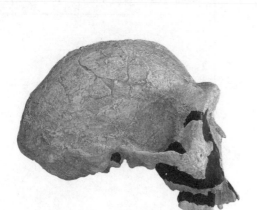

KMN-ER 3733 skull from Koobi Fora region to the east of Lake Turkana, Kenya

Hominin Evolution

Homo erectus

Homo erectus is reserved for the later Asian forms (shown here), with dates ranging from 1 million to 300 000 years ago. Thought to be the first humans to venture out of Africa (but see *ergaster*). They differ from *ergaster* by having skulls that were strongly buttressed with ridges of bone, skull walls greatly thickened, and larger brain volumes (range: 1000 to 1250 cc). These simple hunter-gatherers used stone tools and fire. Nicknames: Java Man, Peking Man, Solo Man.

Artist's reconstruction

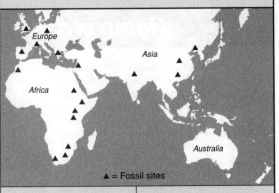

▲ = Fossil sites

Years:		Brain size:	cc
Height:	m	Weight:	kg
Diet:			
Geographic distribution:			
Additional notes:			

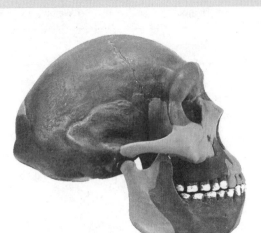

Homo erectus pekinenis or 'Peking Man' from Zhoukoudian Cave, near Peking in China

Archaic *Homo sapiens*

About 300 000 years ago transitional **archaic** forms appear in increasing variation and numbers. Possessing both modern and *Homo erectus*-like features, they were probably 'experimental prototypes' responding to diverse regional selection pressures. They had brain sizes ranging from 1100 to 1400 cc. Some researchers refer to this variable collection of early (archaic) *Homo sapiens* as a separate species: *Homo heidelbergensis*. Nicknames: Rhodesia Man, Steinheim Man, Swanscombe Man, Heidelberg Man.

Artist's reconstruction

▲ = Fossil sites

Years:		Brain size:	cc
Height:	m	Weight:	kg
Diet:			
Geographic distribution:			
Additional notes:			

Homo sapiens rhodesiensis or 'Rhodesia Man', from Broken Hill Mine, Kabwe in Zambia

Homo antecessor

Homo antecessor, a highly controversial species, existed between 780 000 and 625 000 years ago. Discoveries at the Gran Dolina site in the Sierra de Atapuerca, Spain, make these the earliest known European hominin specimens. Fossils consist of nearly 80 postcranial, cranial, facial, and mandibular bones as well as teeth of at least six individuals. *H. antecessor* ('pioneer') shows a mixture of primitive and modern traits, with an especially modern-looking midface.

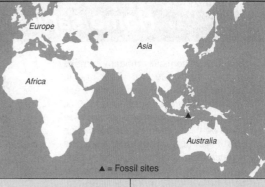

▲ = Fossil sites

Years:		Brain size:	cc
Height:	m	Weight:	kg
Diet:			
Geographic distribution:			
Additional notes:			

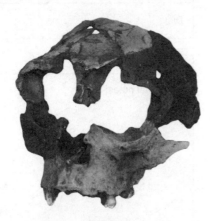

Homo antecessor from Gran Dolina, Sima de los Huesos in the Sierra de Atapuerca, Spain

Homo floresiensis

Homo floresiensis was discovered in 2003 at Liang Bua, a limestone rock shelter on the remote island of Flores, Indonesia. Just one metre tall and with a chimp-size brain, these miniature humans lived 13 000 - 95 000 ya, which means they would have lived at the same time as modern humans. The remains of up to seven individuals have been found alongside pebble tools and the remains of extinct pygmy elephants and giant rodents, and the Komodo dragons, which still live today.

▲ = Fossil sites

Years:		Brain size:	cc
Height:	m	Weight:	kg
Diet:			
Geographic distribution:			
Additional notes:			

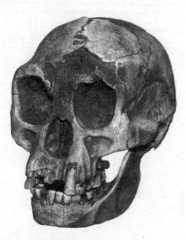

Homo floresiensis, or the 'Hobbit', from the island of Flores, in Indonesia

Hominin Evolution

Homo neanderthalensis

Neanderthals existed between 230 000 and 28 000 years ago. The average brain size was 1450 cc, larger than that of modern humans. They had short, squat, cold-adapted bodies with thick and heavy bones. Men were about 1.7 m tall. Recent research suggests that the Neanderthals split off from the line to modern humans nearly 600 000 years ago. For this reason they are no longer classified as a subspecies of *Homo sapiens* (previously *Homo sapiens neanderthalensis*).

Artist's reconstruction

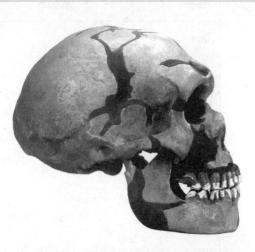

La Ferrassie Skull, Le Bugue, Dordogne Valley in France

▲ = Fossil sites

Years:		Brain size:	cc
Height:	m	Weight:	kg
Diet:			
Geographic distribution:			
Additional notes:			

Homo sapiens

The first anatomically modern humans appear about 160 000 years ago in southern African and the Middle East. The average brain volume was 1350 cc and the skeleton was very gracile. They underwent a sudden cultural revolution about 40 000 years ago, with the appearance of Cro-Magnon culture. Using a wider range of materials, their tools kits became markedly more sophisticated. They were skilled hunters, tool-makers and artists (cave art and music).

Artist's reconstruction

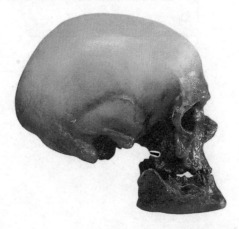

'Cro-Magnon Man', from Cro-Magnon, Dordogne Valley in France

▲ = Fossil sites

Years:		Brain size:	cc
Height:	m	Weight:	kg
Diet:			
Geographic distribution:			
Additional notes:			

The Origin of Modern Humans

There is great debate over the origins of "anatomically modern" humans, i.e. the first emergence of *Homo sapiens*. The two main contesting theories are called **multiregional** and **replacement** hypotheses, with the **assimilation** model (not shown) being a compromise between the two. The "moderns" lack some of the features characteristic of earlier, archaic hominins, such as the protruding snout and heavy brow ridges. The modern skulls have an essentially flat face, are globular (rather than elongated), and have a more nearly vertical forehead. The face is narrower and smaller, and the jaw has a protruding chin. The rest of the skeleton is less robust.

Multiregional Hypothesis

Advocates: Milford Wolpoff, University of Michigan
Alan Thorne, Australian National University

Based largely on the fossil evidence and the anatomical characteristics of modern populations, 'multiregional evolution' traces all modern populations back at least 1 million years to when early humans (*Homo erectus*) first left Africa. Modern *Homo sapiens* emerged gradually throughout the world, and as the populations dispersed, they remained in 'genetic contact'. This gene flow between neighbouring populations ensured that the general 'modern human blueprint' was adopted by all. This limited gene flow still allowed for slight anatomical differences to be retained or develop in the regional populations. Wolpoff and Thorne who are advocates of this theory maintain that the mitochondrial DNA data can be interpreted in a way that supports the multi-regional model.

See *Scientific American* SPECIAL ISSUE: New Look at Human Evolution, Vol. 13(2), 2003, pp. 46-61, for two excellent articles on each of these models.

Replacement Hypothesis

Advocates: Christopher Stringer, Natural History Museum in London
The late Alan C. Wilson

Also known as the "Out of Africa Hypothesis" and "Eve Hypothesis". This model keeps *Homo sapiens* as a separate species and states that modern humans evolved from archaics in only one location, Africa, and then spread, replacing the archaic populations when they came in contact. The extinction of these regional archaic populations occurred because the modern humans were better adapted. In support of this theory, the late Allan C. Wilson and colleagues carried out genetic studies on modern endemic human populations. They concluded that the evolutionary record of mitochondrial DNA could be traced back to a single female who lived in Africa some 200 000 years ago. This woman, real or hypothetical, has been dubbed 'Eve' by Wilson and his team. By implication, this theory maintains that all modern descendants contain mitochondrial DNA that can be traced directly back to Eve.

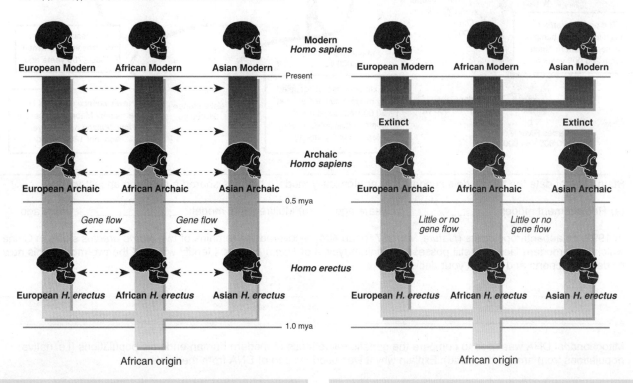

Predictions made by this model

1. Fossils that show the change from one stage to the next in all geographic regions (transitional forms).

2. Modern traits should appear in the fossil record somewhat simultaneously all over the Old World range of Archaic *Homo sapiens*.

3. Today's modern "racial" traits characteristic of a particular region can be traced back to ancient forms in that region.

4. The human species today should have a high degree of genetic diversity since it is an old species with distinct populations that have had a lot of time to accumulate genetic differences.

5. The amount of genetic variation within each modern human group is about the same since they have all been evolving together.

Predictions made by this model

1. Transitional forms would be found in only one place (in this case Africa) which is the area of origin for modern humans.

2. Modern traits should appear first in one location (Africa) and then later elsewhere as the modern population spread to other parts of the Old World.

3. Modern and archaic populations should overlap in time outside the area that moderns originated (the process of replacement would not be instantaneous).

4. Humans today should have relatively little genetic diversity since the species is young.

5. Today's modern populations should differ in the amount of genetic variation, the most diversity being found in the region where moderns first evolved (this would have been the oldest group and therefore the one that had the most time for genetic variation to accumulate).

Source: Michael A. Park, Biological Anthropology, Mayfield Publishing, 1996

Code: DA 3

The map below shows a probable origin and dispersal of modern humans throughout the world. An African origin is almost certain, with south eastern Africa being the most likely region. The dispersal was affected at crucial stages by the presence or absence of 'land bridges' formed during the drop in sea level that occurs with the onset of ice ages. The late development of boating and rafting technology slowed dispersal into Australia and the Pacific. New Zealand was one of the last places on Earth to be populated. (On the map, **ya** = years ago.)

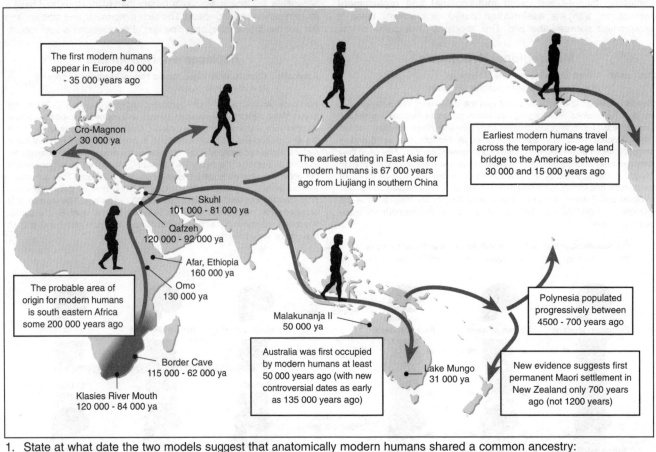

The first modern humans appear in Europe 40 000 - 35 000 years ago

Cro-Magnon 30 000 ya

The earliest dating in East Asia for modern humans is 67 000 years ago from Liujiang in southern China

Earliest modern humans travel across the temporary ice-age land bridge to the Americas between 30 000 and 15 000 years ago

Skuhl 101 000 - 81 000 ya

Qafzeh 120 000 - 92 000 ya

Afar, Ethiopia 160 000 ya

Omo 130 000 ya

The probable area of origin for modern humans is south eastern Africa some 200 000 years ago

Polynesia populated progressively between 4500 - 700 years ago

Malakunanja II 50 000 ya

Border Cave 115 000 - 62 000 ya

Australia was first occupied by modern humans at least 50 000 years ago (with new controversial dates as early as 135 000 years ago)

Lake Mungo 31 000 ya

New evidence suggests first permanent Maori settlement in New Zealand only 700 years ago (not 1200 years)

Klasies River Mouth 120 000 - 84 000 ya

1. State at what date the two models suggest that anatomically modern humans shared a common ancestry:

   (a) Replacement model: _____ years ago   (b) Multiregional model: _____ years ago

2. In 1992, palaeoanthropologists (*Nature*, Vol. 357, page 404) recovered the remains of two *Homo erectus* skulls in China, which have modern faces but still possess a cranium typical of *Homo erectus*. Identify which of the two models this new evidence supports and explain your decision:

   _____

   _____

3. Mitochondrial DNA was used to compare the genetic relatedness of modern human endemic populations (i.e. native populations from around the world). Explain why it was used instead of DNA from the nucleus:

   _____

   _____

4. Explain the significance of the **gene flow** between early populations of humans in the multiregional model:

   _____

   _____

5. In the out of Africa model, modern humans move out of Africa to populate the rest of the world. Describe the fate of the other human populations already inhabiting these regions, according to this theory:

   _____

   _____

6. Discuss the implications of an early (135 000 ya) arrival of modern humans into Australia: _____

   _____

   _____

Distinguishing Features of Hominins

The data below provide you with lists of features that distinguish the many hominin species from each other. In your reading, the 'known dates' provided may vary from those given below, mainly due to varying interpretations on the dating of sites by different researchers. Some early hominin species, including various australopithecines and *Kenyanthropus*, are not listed.

Distinguishing Features of Early Human Species

	Homo habilis (small)	Homo habilis (large)	Homo erectus	Archaic Homo sapiens	Homo neanderthalensis	Early Homo sapiens
Other name	None	*Homo rudolfensis*	*Homo ergaster* for older African forms	*Homo heidelbergensis*	The Neanderthals	Early anatomically modern humans
Known date (years ago)	2 – 1.6 million	2.4 – 1.6 million	1.8 – 0.3 million	400 000 - 100 000	150 000 - 30 000	160 000 - 60 000
Brain size	500-650 cc	600-800 cc	750-1250 cc	1100-1400 cc	1200-1750 cc	1200-1700 cc
Height	1.0 m	*c.* 1.5 m	1.3 - 1.5 m	?	1.5-1.7 m	1.6-1.85 m
Physique	Relatively long arms	Robust but 'human' skeleton	Robust but 'human' skeleton	Robust but 'human' skeleton	Robust but 'human' skeleton, adapted for cold	Modern skeleton possibly adapted for warmth
Skull shape	Small face with developed nose	Larger, flatter face	Flat, thick skull with sagittal 'keel' and large brow ridge	Higher cranium, face less protruding	Reduced brow ridge, midface projection, long low skull	Small or no brow ridge, shorter and higher skull
Teeth and jaws	Smaller, narrow molars; thinner jaw	Large narrow molars; robust jaw	Smaller teeth than *H. habilis*, robust jaw in larger individuals	Similar to *H. erectus* but smaller teeth	Similar to Archaic *H. sapiens*; except for incisors, smaller teeth	Teeth may be smaller; shorter jaws than Neanderthals; chin developed
Geographical distribution	Eastern, and possibly Southern Africa	Eastern Africa possibly? western Asia (Rep. Georgia)	Africa, Asia, Indonesia, and possibly Europe	Africa, Asia and Europe	Europe and western Asia	Africa and western Asia

Distinguishing Features of Early Hominins

	Orrorin tugenensis	Ardipithecus ramidus	Australopithecus anamensis	Australopithecus afarensis	Australopithecus africanus	Paranthropus robustus
Other name	"Millennium Man"	Two subspecies: *ramidus* & *kadabba*	None	None	None	*Australopithecus robustus*
Known date (years ago)	6.0 million	4.4 – 5.8 million	4.2 – 3.9 million	3.9 – 2.5 million	~3.0 – 2.3 million	2.2 – 1.5 million
Brain size	? cc	? cc	? cc	400 – 500 cc	400 – 500 cc	530 cc
Height	? m	*c.* 1.22 m	? m	1.07 – 1.52 m	1.1 – 1.4 m	1.1 – 1.3 m
Physique	Possibly bipedal forest dweller. Little else known	Possibly bipedal forest dweller. Little else known	Partial leg bones strongly suggest bipedalism; humerus extremely humanlike	Light build. Some apelike features: relatively long arms, curved fingers/toes, sexual dimorphism	Light build. Probably long arms, more 'human' features, probably less sexual dimorphism	Heavy build. Relatively long arms. Moderate sexual dimorphism
Skull shape	Not yet described	Foramen magnum more forward than apes	Primitive features in the skull, possibly apelike	Apelike face, low forehead, bony brow ridge, flat nose, no chin	Brow ridges less prominent; higher forehead and shorter face	Long, broad, flat face; crest on top of skull; moderate facial buttressing
Teeth and jaws	Not yet described	Teeth are intermediate between those of *A. afarensis* and earlier apes. Smaller, narrow molars; thinner jaw	Very similar to those of older fossil apes, but canines vertical; teeth have thicker tooth enamel like in humans	Human-like teeth, canines smaller than apes, larger than humans. Jaw shape half way between an ape's and human.	Teeth and jaws much larger than in humans; tooth row fully parabolic like humans; canine teeth further reduced	Very thick jaws; small incisors and canines; large molar-like premolars; very large molars
Geographical distribution	Eastern Africa	Eastern Africa	Eastern Africa	Eastern Africa	Southern Africa	Southern Africa

Hominin Evolution

Bipedalism and Nakedness

The first major step in the development of humans as a distinct group from apes was their ability to adopt a habitually upright stance. Closely linked to this shift in the mode of locomotion was the reduction in body hair (we are the only 'naked ape'). A number of selection pressures for hair reduction are suggested below. Some experts have suggested that bipedalism and hair reduction were evolutionary responses to the changing climate of East Africa about 7-5 million years ago (see below). Numerous reasons have been offered for the development of bipedalism in hominins. The need to adjust to a newly emerging habitat was probably important, although the earliest hominins may still have been forest dwellers. Gone is the old image of bipedalism evolving as a result of a move into the savannah (open grassland). Current theories must explain how bipedalism

may have emerged in a more forested environment. Animal and plant fossils at hominin sites have indicated that the earliest bipedal hominins frequented a variety of habitats, including **open woodland**, **gallery forest**, **closed woodland**, and **savannah**. The considerable number of early hominin species living at the same time may reflect local adaptations to different environments. As forests receded in Africa due to the drying of the global climate, vast areas of savannah became established. Many primate species were not able to adjust to the loss of trees since they were primarily adapted to tree-dwelling. These primates retreated with their shrinking range. Others met the challenge by successfully adopting new behaviour patterns that resulted in anatomical adaptations. Baboons and the hominins succeeded in this, while others failed and became extinct.

Hair Reduction

Retention of head hair

Hair on the head (and to a lesser extent the shoulders) has been retained to reflect and radiate heat before it reaches the skin of the exposed part of the body.

Parasite control

With the reduction in body hair, control of parasites such as fleas, ticks and lice would have been improved. This became particularly important when early hominids began to use a 'home base' rather than continually wandering. Parasites such as fleas need to complete their life cycle at a single location so that hatching eggs can reinfect their host.

Lice

Ticks

Fleas

Thermoregulation

Shorter, finer hairs (not hair loss) in early hominids has allowed greater heat loss by radiating from the skin surface. Well developed sweat glands allow us to lose heat at an astounding 700 watts m² of skin (a capacity not approached by any other mammal).

Bipedalism

Seeing over the grass

Being upright may have helped to spot predators or locate carcasses at a distance.

Carrying food

The capacity to carry food away from a kill site or growing site to a position of safety would have had great survival advantage.

Carrying offspring

Walking upright enabled early hominins to carry their offspring while following the large game herds of the savannah on long seasonal migrations.

Hold tools and weapons

Tool use was probably a consequence of bipedalism, rather than a cause. Upright walking appears to have been established well before the development of hunting in early hominins.

Thermoregulation

Advantages of the upright walking are:
1. 60% less surface area presented to the sun at midday.
2. Greater air flow across the body when it is lifted higher off the ground. The air higher up is also less humid and cooler.

Efficient locomotion

With the change in habitat resulting from the cooling and drying of the Late Miocene a more efficient means of moving across the growing expanses of savannah was required. Bipedalism provides an energy efficient method that favours low speed, long distance movement – walking.

The changing climate of East Africa in the Miocene

As the climate and habitat changed, early hominins in the wooded savannah would have been forced to move across open ground to exploit their food resources amongst the trees. They would also have been under pressure to experiment with new food resources.

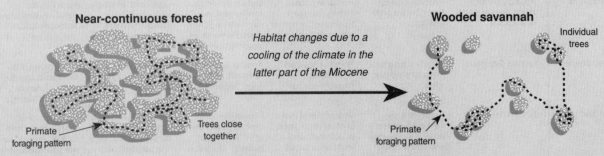

Near-continuous forest

Habitat changes due to a cooling of the climate in the latter part of the Miocene

Wooded savannah

Individual trees

Primate foraging pattern

Trees close together

Primate foraging pattern

Pre-hominins foraged for food in nearly continuous forest; food resources were readily available. A near completely arboreal life was possible with only the occasional need to move on the ground.

By the late Miocene, some early hominins were faced with a very different habitat of widely separated trees. They were probably forced to experiment with savannah food resources.

Adaptations for Bipedalism

Important modifications in the skeleton are associated with the move to bipedal locomotion in early hominins. The skeleton opposite is an example of an early bipedal hominin. It is a reconstruction of 'Lucy' (*Australopithecus afarensis*) dated at about 3 million years ago. While Lucy still possessed some ape-like characteristics, such as curved toes, she was a fully-bipedal hominin with all the modern adaptations associated with modern human walking. Lucy was a small individual, only 1.1 metres tall, about the height of a 5-6 year old child. Although there is no doubt that Lucy was habitually bipedal, a number of skeletal features suggest that tree climbing was still an important part of this hominin's niche. Such activities may have been associated with escape from predators, obtaining a secure sleeping place, and foraging for foods found in trees. The features that point to a link with **arboreal** (tree-dwelling) locomotion are indicated on the left of the diagram. *A. afarensis* forms an important link between the quadrupedal locomotion of apes and bipedalism in hominins.

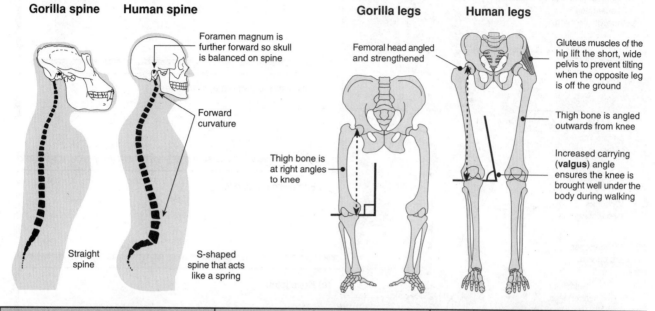

Gorilla spine **Human spine**

Foramen magnum is further forward so skull is balanced on spine

Forward curvature

Straight spine

S-shaped spine that acts like a spring

Gorilla legs **Human legs**

Femoral head angled and strengthened

Gluteus muscles of the hip lift the short, wide pelvis to prevent tilting when the opposite leg is off the ground

Thigh bone is angled outwards from knee

Thigh bone is at right angles to knee

Increased carrying (**valgus**) angle ensures the knee is brought well under the body during walking

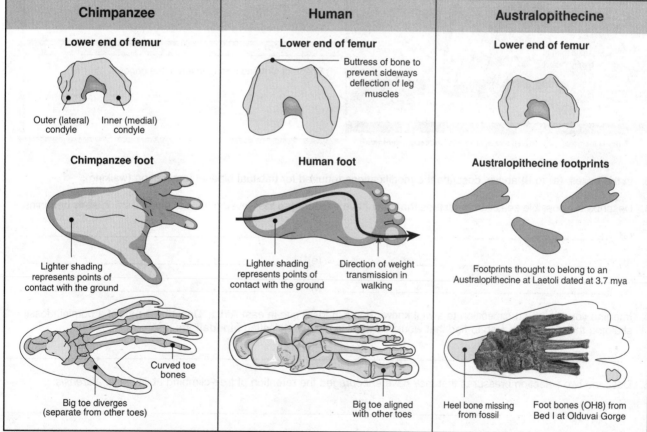

Chimpanzee	Human	Australopithecine
Lower end of femur	**Lower end of femur**	**Lower end of femur**
Outer (lateral) condyle Inner (medial) condyle	Buttress of bone to prevent sideways deflection of leg muscles	
Chimpanzee foot	**Human foot**	**Australopithecine footprints**
Lighter shading represents points of contact with the ground	Lighter shading represents points of contact with the ground Direction of weight transmission in walking	Footprints thought to belong to an Australopithecine at Laetoli dated at 3.7 mya
Big toe diverges (separate from other toes) Curved toe bones	Big toe aligned with other toes	Heel bone missing from fossil Foot bones (OH8) from Bed I at Olduvai Gorge

1. Referring to the diagram above, describe whether each of the **australopithecine fossils** compare more closely to the *chimpanzee* or *human* examples (i.e. to which do they bear the closest resemblance):

   (a) Lower end of femur: _____

   (b) Footprints: _____

   (c) Foot skeleton: _____

Hominin Evolution

Code: RA 3

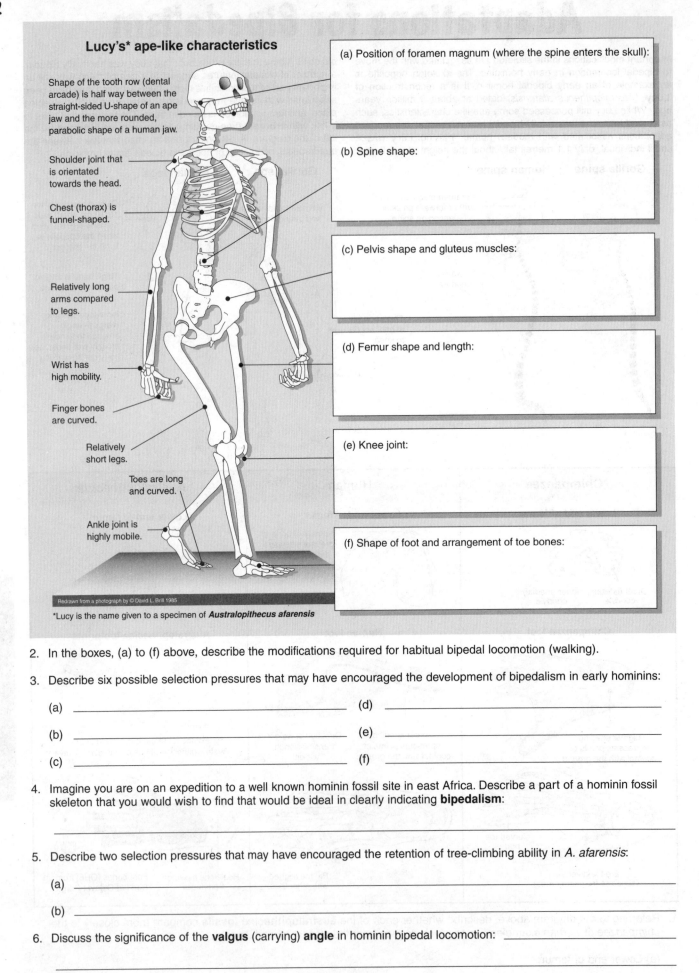

Lucy's* ape-like characteristics

Shape of the tooth row (dental arcade) is half way between the straight-sided U-shape of an ape jaw and the more rounded, parabolic shape of a human jaw.

Shoulder joint that is orientated towards the head.

Chest (thorax) is funnel-shaped.

Relatively long arms compared to legs.

Wrist has high mobility.

Finger bones are curved.

Relatively short legs.

Toes are long and curved.

Ankle joint is highly mobile.

Redrawn from a photograph by © David L. Brill 1985

*Lucy is the name given to a specimen of *Australopithecus afarensis*

(a) Position of foramen magnum (where the spine enters the skull):

(b) Spine shape:

(c) Pelvis shape and gluteus muscles:

(d) Femur shape and length:

(e) Knee joint:

(f) Shape of foot and arrangement of toe bones:

2. In the boxes, (a) to (f) above, describe the modifications required for habitual bipedal locomotion (walking).

3. Describe six possible selection pressures that may have encouraged the development of bipedalism in early hominins:

(a) _____ (d) _____

(b) _____ (e) _____

(c) _____ (f) _____

4. Imagine you are on an expedition to a well known hominin fossil site in east Africa. Describe a part of a hominin fossil skeleton that you would wish to find that would be ideal in clearly indicating **bipedalism**:

_____

5. Describe two selection pressures that may have encouraged the retention of tree-climbing ability in *A. afarensis*:

(a) _____

(b) _____

6. Discuss the significance of the **valgus** (carrying) **angle** in hominin bipedal locomotion: _____

_____

_____

The Development of Intelligence

The human brain is an extraordinary organ and is responsible for our unique human behavioural qualities. Although it makes up just 2% of our body weight, it demands about 20% of the body's metabolic energy at rest. This makes the brain an expensive organ to maintain. The selection pressures for increased brain size must have been considerable for additional energy to be made available. The normal human adult brain averages around 1330 cc, but ranges in size between 1000 and 2000 cc. The modern brain contains as many as 10 000 million nerve cells, each of which has thousands of synaptic connections with other nerve cells. But intelligence is not just a function of **brain size**. There are large mammals, such as elephants and whales, with brain volumes greater than ourselves and yet they are not considered to be as intelligent. It appears that what is more important is the **relative brain size** (brain size relative to body size). Modern humans have a brain volume three times larger than that predicted for an average monkey or ape with our body size. Another important factor is the way in which the brain is organised. Apart from the highly developed cerebrum, two areas of the brain associated with communicaton have also become highly developed in modern humans: **Broca's area**, concerned with speech, and **Wernicke's area**, concerned with comprehension of language.

Growth in Brain size in Humans and Chimpanzees

In most primates, including chimpanzees, brain growth, relative to body size, slows markedly after birth while body growth continues. In human infants, the slowing of brain growth does not occur until more than a year after birth, which results in larger brain masses for humans than for chimpanzees at any given age (or body weight).

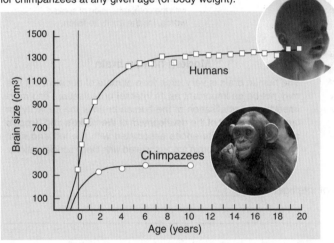

Brain Volume for Hominin Species

This table provides a generalised summary of the changes in estimated brain volume recorded from the fossil remains of hominins. The dates for each species are generally the middle of their time range for long-lived species or at the beginning of their time range for short-lived species.

Hominin species	Years ago (mya)	Average brain Volume (cm³)
Australopithecus afarensis	3.5	440
Australopithecus africanus	2.5	450
Paranthropus robustus	2.0	520
Paranthropus boisei	1.5	515
Homo rudolfensis	2.0	700
Homo habilis	1.8	575
Homo ergaster	1.8	800
Homo erectus	0.5	1100
Homo heidelbergensis	0.2	1250
Homo neanderthalensis	0.05	1550
Homo floresiensis	0.095*	380
Early *Homo sapiens*	0.08	1450

* *H. floresiensis* may have lived as recently as 13 000 ya

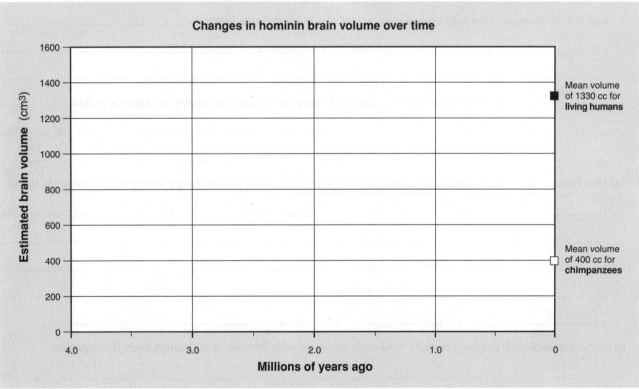

Changes in hominin brain volume over time

Mean volume of 1330 cc for **living humans**

Mean volume of 400 cc for **chimpanzees**

1. Plot the data in the table on the estimated *Brain Volume for Hominin Species* (above) onto the graph provided.

2. There were two 'bursts' (sudden increases) of brain expansion during human evolution. **Indicate on the graph** you have plotted where you think these two events occurred.

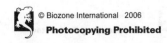
Hominin Evolution

Code: RDA 3

Brain Size vs Body Height in Hominins

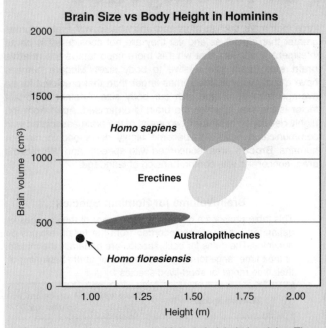

Brain size can be correlated with body height in hominins. Three distinct clusters emerge, indicating three phases of evolutionary development. *Homo floresiensis*, found on the Indonesian island of Flores, clearly falls outside these clusters. Its brain size to body size ratio is similar to that of the Australopithecines, but key aspects of its morphology, such as its small canine teeth and organisation of the brain, identify it as *Homo*. In addition, the Flores finds were associated with relatively advanced stone tools.

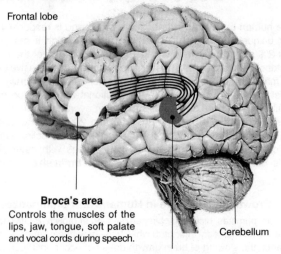

Frontal lobe

Broca's area
Controls the muscles of the lips, jaw, tongue, soft palate and vocal cords during speech.

Cerebellum

Wernicke's area
The area of the brain concerned with the comprehension of spoken words, i.e. the ability to listen.

Modern human brain

The human brain is very large for a primate of our size, but this may not be as important as its internal organisation. The most important specialisation of the human brain is the capacity for language: a result of the development of **Wernicke's** and **Broca's areas**. Specific differences associated with the left and right hemispheres of the brain are associated with these specialisations.

3. Explain why brain volume alone is not a reliable indicator of intelligence: _____

_____

_____

4. Explain the significance of the high energy requirement of a relatively large brain: _____

_____

_____

5. Comment on the significance of the brain/body size growth curve in humans compared with other primates:

_____

_____

6. (a) With respect to stature and brain size, comment on the position of *Homo floresiensis* with respect to other hominins:

_____

_____

(b) Comment on the significance of the Flores finds: _____

_____

_____

7. Describe a **selection pressure** that might have been acting on early humans to encourage brain development:

_____

_____

_____

_____

Dating Fossils

Fossils are rarely able to be dated directly. In general, it is the rocks in which they are found that are dated. The exception is radiocarbon dating, which can directly measure the age of the organic matter in a sample. Dating usually begins with an attempt to order past events in a rock profile, and to relate the fossils to datable rock layers. Many techniques for measuring the age of rocks and minerals have been established. In the early days of developing these techniques there were problems in producing dependable results, but the methods have been much refined and often now provide dates with a high degree of certainty. Multiple dating methods may be applied to samples, providing cross-referencing, which gives further confidence in a given date. Dating methods can be grouped into two categories: those that rely on the gradual radioactive decay of an element (e.g. **radiocarbon**, **potassium-argon**, **fission track**); and those that use other methods (e.g. **tree-rings**, **palaeomagnetism**).

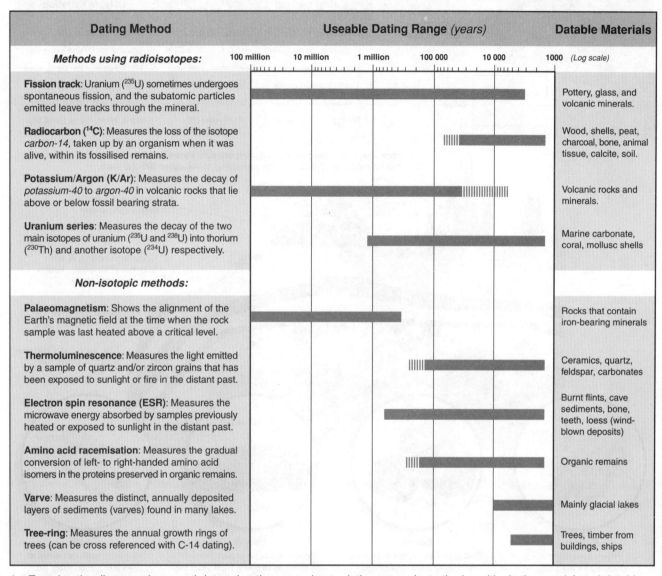

Dating Method	Useable Dating Range *(years)*	Datable Materials
Methods using radioisotopes:	100 million 10 million 1 million 100 000 10 000 1000 *(Log scale)*	
Fission track: Uranium (^{235}U) sometimes undergoes spontaneous fission, and the subatomic particles emitted leave tracks through the mineral.		Pottery, glass, and volcanic minerals.
Radiocarbon (^{14}C): Measures the loss of the isotope *carbon-14,* taken up by an organism when it was alive, within its fossilised remains.		Wood, shells, peat, charcoal, bone, animal tissue, calcite, soil.
Potassium/Argon (K/Ar): Measures the decay of *potassium-40* to *argon-40* in volcanic rocks that lie above or below fossil bearing strata.		Volcanic rocks and minerals.
Uranium series: Measures the decay of the two main isotopes of uranium (^{235}U and ^{238}U) into thorium (^{230}Th) and another isotope (^{234}U) respectively.		Marine carbonate, coral, mollusc shells
Non-isotopic methods:		
Palaeomagnetism: Shows the alignment of the Earth's magnetic field at the time when the rock sample was last heated above a critical level.		Rocks that contain iron-bearing minerals
Thermoluminescence: Measures the light emitted by a sample of quartz and/or zircon grains that has been exposed to sunlight or fire in the distant past.		Ceramics, quartz, feldspar, carbonates
Electron spin resonance (ESR): Measures the microwave energy absorbed by samples previously heated or exposed to sunlight in the distant past.		Burnt flints, cave sediments, bone, teeth, loess (wind-blown deposits)
Amino acid racemisation: Measures the gradual conversion of left- to right-handed amino acid isomers in the proteins preserved in organic remains.		Organic remains
Varve: Measures the distinct, annually deposited layers of sediments (varves) found in many lakes.		Mainly glacial lakes
Tree-ring: Measures the annual growth rings of trees (can be cross referenced with C-14 dating).		Trees, timber from buildings, ships

Hominin Evolution

1. Examine the diagram above and determine the approximate dating range (note the logarithmic time scale) and datable materials for each of the methods listed below:

	Dating Range	Datable Materials
(a) Potassium-argon method:	_____	_____
(b) Radiocarbon method:	_____	_____
(c) Tree-ring method:	_____	_____
(d) Thermoluminescence:	_____	_____

2. When the date of a sample has been determined, it is common practice to express it in the following manner:
    Example: **1.88 ± 0.02** million years old. Explain what the **± 0.02** means in this case:

    _____

3. Suggest a possible source of error that could account for an incorrect dating measurement using a radioisotope method:

    _____

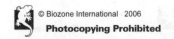

Code: RDA 2

Dating a Fossil Site

The diagram below shows a rock shelter typical of those found in the Dordogne Valley of Southwest France. Such shelters have yielded a rich source of Neanderthal and modern human remains. It illustrates the way in which hominin activity is revealed at archaeological excavations. Occupation sites included shallow caves or rocky overhangs of limestone. The floors of these caves accumulated the debris of natural rockfalls, together with the detritus of human occupation at various layers, called **occupation horizons**. A wide array of techniques can be used for dating, some of which show a high degree of reliability (see the table below). The use of several appropriate techniques to date material improves the reliability of the date determined.

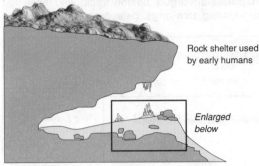

Rock shelter used by early humans

Enlarged below

Dating method	Dating range (years ago)	Datable materials
Radiocarbon (^{14}C)	1000 - 50 000+	Bone, shell, charcoal
Potassium-argon (K/Ar)	10 000 - 100 million	Volcanic rocks and minerals
Uranium series decay	less than 1 million	Marine carbonate, coral, shell
Thermoluminescence	less than 200 000	Ceramics (burnt clay)
Fission track	1000 - 100 million	Volcanic rock, glass, pottery
Electron spin resonance	2000 - 500 000	Bone, teeth, loess, burnt flint

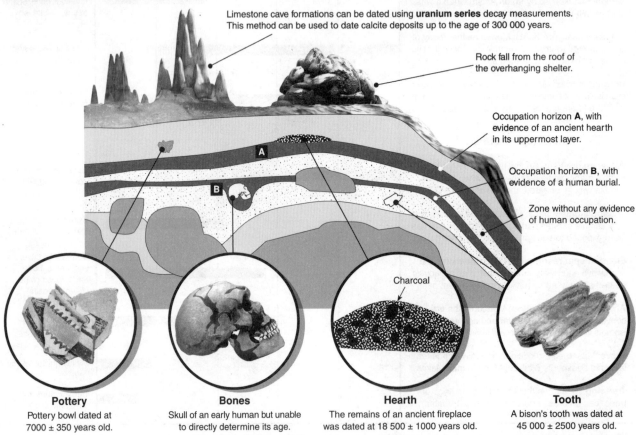

Limestone cave formations can be dated using **uranium series** decay measurements. This method can be used to date calcite deposits up to the age of 300 000 years.

Rock fall from the roof of the overhanging shelter.

Occupation horizon **A**, with evidence of an ancient hearth in its uppermost layer.

Occupation horizon **B**, with evidence of a human burial.

Zone without any evidence of human occupation.

Charcoal

Pottery
Pottery bowl dated at 7000 ± 350 years old.

Bones
Skull of an early human but unable to directly determine its age.

Hearth
The remains of an ancient fireplace was dated at 18 500 ± 1000 years old.

Tooth
A bison's tooth was dated at 45 000 ± 2500 years old.

1. Discuss the significance of **occupation horizons**: _____

_____

_____

_____

2. Determine the approximate date range for the items below: (Hint: take into account layers/artifacts with known dates)

   (a) The skull at point B: _____

   (b) Occupation horizon A: _____

3. Name the dating methods that could have been used to date each of the following, at the site above:

   (a) Pottery bowl: _____   (c) Hearth: _____

   (b) Skull: _____   (d) Tooth: _____

Interpreting Fossil Sites

Human skull

Charcoal fragments (possible evidence of fire use and excellent for rediocarbon dating).

Bones from a large mammal with evidence of butchering (cut and scrape marks from stone tools). These provide information on the past ecology and environment of the hominins in question.

Excavation through rock strata (layers). The individual layers can be dated using both chronometric (absolute) and relative dating methods.

Stone tools

Photo: RA

istock

Searching for ancient human remains, including the evidence of culture, is the work of **palaeoanthropologists**. Organic materials, such as bones and teeth, are examined and analysed by physical anthropologists, while cultural materials, such as tools, weapons, shelters, and artworks, are examined by archaeologists. Both these disciplines, **palaeoanthropology** and **archaeology**, are closely associated with other scientific disciplines, including **geochemistry** (for **chronometric dates**), **geology** (for reconstructions of past physical landscapes), and **palaeontology** (for knowledge of the past species assemblages).

The reconstruction of a **dig site**, pictured above, illustrates some of the features that may be present at a site of hominin activity. Naturally, the type of information recovered from a site will depend on several factors, including the original nature of the site and its contents, the past and recent site environment, and earlier disturbance by people or animals. During its period of occupation, a site represents an interplay between additive and subtractive processes; building vs destruction, growth vs decay. Organic matter decays, and other features of the site, such as tools, can be disarranged, weathered, or broken down. The archaeologists goal is to maximise the recovery of information, and recent trends have been to excavate and process artifacts immediately, and sometimes to leave part of the site intact so that future work, perhaps involving better methodologies, is still possible.

4. Explain why palaeoanthropologists date and interpret all of the remains at a particular site of interest (e.g. animal bones, pollen, and vegetation, as well as hominin remains):

_____

_____

_____

_____

5. Discuss the importance of involving several scientific disciplines when interpreting a site of hominin activity:

_____

_____

_____

_____

_____

Hominin Evolution

Cultural Evolution

Describing human cultural evolution

Tool use (stone, wood, bone, iron and bronze), fire, food-gathering, shelter, clothing, abstract thought and domestication of plants and animals.

Learning Objectives

☐ 1. Compile your own glossary from the **KEY WORDS** displayed in **bold type** in the learning objectives below.

Cultural Evolution *(pages 53-54, 59-70)*

☐ 2. Distinguish between **cultural evolution** and **biological (physical) evolution** and describe the relative importance of each in human evolution. Recognise that, by necessity, they are studied together, rather than in isolation. Identify and discuss general trends in hominin cultural development, including tool use, spiritual practices, and the development of abstract thought.

☐ 3. Describe the features of the various **Palaeolithic** (or Old Stone Age) tool cultures, including the hominin species associated with each. Identify the advancements in tool technology made at each stage:
   (a) Oldowan (pebble) tool culture
   (b) Acheulian tool culture
   (c) Mousterian tool culture
   (d) Upper Palaeolithic tool culture

☐ 4. Describe the development of successive cultures, i.e. **Mesolithic** (or Middle Stone Age), **Chalcolithic** (Copper-Stone) Age, **Neolithic** (or New Stone Age), and **Bronze Age** cultures. Identify the advancements in tool technology made at each stage and the particular features characteristic of each.

☐ 5. Discuss how the evolution of more complex behaviours (the use of tools, fire, clothing, beliefs and spirituality) were aided by a growing capacity for learning and communication. Provide examples of the evidence for increasingly sophisticated cultural development towards the end of the Upper Palaeolithic period.

Future Evolution of Humans *(pages 71-72)*

☐ 6. EXTENSION: Debate (in groups or as a class) the likely selection pressures currently acting on the human gene pool. Discuss likely future trends in human biological and cultural evolution, giving reasons for your ideas. Analyse the possible effects on human evolution of increased population mobility, modern medicine, and genetic engineering. Explain how the development of these behaviours has been aided by a growing capacity for learning and communication.

Supplementary Texts

See page 7 for additional details of these texts:
■ Coppens, Y. 2004. **Human Origins: The Story of our Species**, reading as required.
■ Jones *et al.* 1992. **The Cambridge Encyclopedia of Human Evolution**, (CUP), chapters 9-10.
■ Lynch, J. and Barrett, L., 2003. **Walking With Cavemen (BBC)**, as required.
■ Mai, L.I. *et al*, 2005. **The Cambridge Dictionary of Human Biology & Evolution**, (CUP), as required as reference.

Periodicals

See page 7 for details of publishers of periodicals:

STUDENT'S REFERENCE

■ **Fired up** New Scientist, 20 May 2000, pp. 30-34. *New evidence suggests that hominins may have been using fire for longer than previously thought.*

■ **France's Magical Cave Art: Chauvet Cave** National Geographic, 200(2) August 2001, pp. 104-121. *Detailed description of a cave decorated with fabulous prehistoric artwork dated at 35 000 years.*

■ **Just Like Your Mother Taught You** New Scientist, 1 April 2006, pp. 42-45. *The human talent for learning cultural traditions and copying.*

■ **Evolution and Us** New Scientist, 11 March 2006, pp. 30-33. *Recent research indicates that evolution in humans is still taking place and did not come to a halt 50 000 years ago.*

■ **We Believe** New Scientist, 28 January 2006, pp. 30-33. *Religion from an evolutionary biology point of view. What are it's cultural origins and for what purpose did religion evolve?*

■ **A Monumental Collapse?** New Scientist, 29 July 2006, pp. 30-34. *What happened to the civilisation on Easter Island is largely speculation but more recent ecological investigation may shed some light on its downfall.*

TEACHER'S REFERENCE

NEW LOOK AT HUMAN EVOLUTION Scientific American **SPECIAL ISSUE** 13(2) July 2003. *A collection of 12 features (some of which are new and some are updated). Includes:*

■ **Food for Thought** pp. 62-71. *Dietary change as a driving force in human evolution.*

■ **Once Were Cannibals** pp. 86-93. *The practice of cannibalism may be deep-rooted in our history.*

■ **If Humans Were Built to Last** pp. 94-100. *We would look a lot different if evolution had produced the body to function smoothly in youth and old age.*

BECOMING HUMAN Scientific American **SPECIAL ISSUE** 16(2) 2006. *A collection of features (previously published) covering aspects of primate behaviour and evolution. Includes:*

■ **How We Came to be Human** pp. 68-73. *Language and art are hallmarks of humanity.*
■ **The Emergence of Intelligence** pp. 84-92. *Language, foresight and other hallmarks of intelligence are likely to be related to inherent aptitude for certain tasks.*

Internet

See pages 4-5 for details of how to access **Bio Links** from our web site: **www.thebiozone.com** From Bio Links, access sites under the topics: **HUMAN EVOLUTION > Cultural Evolution:** •
Stone age reference collection • Stone pages • The Chauvet cave > **Human Fossil Record:** •
Becoming human • Early human evolution •
Human evolution • The human origins program

Presentation MEDIA to support this topic:

HUMAN EVOLUTION

Cultural Evolution

Natural selection acting on the expression of genes brought about considerable transformations in the anatomy of early humans. In addition, it was possible for ideas and behaviours that were learned to be passed on to offspring. This non-genetic means of adaptation, called **cultural evolution**, further enhanced the success of early humans.

Resulting physical features

The physical features that developed in response to selection pressures of the environment include:

Head balanced on the top of the backbone, instead of held up by large neck muscles. Large brain capable of learning, planning and passing on ideas. Very keen eyesight, capable of judging distances with eyes located high above the ground. Other senses are less well developed. Light but strong jaw, with teeth suitable for varied foods. Backbone slightly curved, allowing upright standing on two legs without getting tired, thus freeing hands, and giving eyes good all round vision. Hands able to grasp and manipulate objects in a very sensitive way. Legs that allowed efficient walking and running on two legs. Flexible ankle, but rigid and arched foot, allowing efficient walking on hard ground.

Environmental forces

Over many millions of years, the evolution of human ancestors has been directed by the forces of natural selection. Environmental forces such as climatic change causing alterations in habitat and food supply, as well as fierce predators, acted on the gene pool.

Climatic change

The climate became drier and the forests which were the homes of the earlier primates gradually disappeared. This not only reduced shelter but also meant that traditional food sources became scarce or disappeared. New food resources had to be experimented with.

Fierce predators

Many large, fierce predators made a ground dwelling lifestyle dangerous. Early humans would have to protect themselves from attack using smart behavioural solutions.

Adopted niche

An opportunist/scavenger that was able to live reasonably successfully on the ground. Able to exploit a number of varied habitats, early humans utilised a range of food resources.

Cultural forces

Because of the unique combination of brain and specialised physical features mentioned above, early humans very gradually began to direct the course of their own evolution. They began to control their environment; to use it to alter their way of life. At first they did this in small ways, and with little effect on other living things. The result was the development of efficient hunters living in organised groups. Their genes had been almost unchanged, but they now lived more comfortably, with a better survival rate, and more time to plan ahead.

Tool making
Tools made by chipping stones, or shaping bones or wood were used in a wide variety of ways. In some cases, the use of tools replaced the need to develop physical features.

Fire making
Fire is a powerful tool. It provided a means of keeping warm in cold periods, deterring predators from a camp site, and driving animals during a hunt. It was also used to cook, allowing difficult to digest food to be eaten more easily.

Shelter and clothing
The earliest shelters were probably natural ones, such as caves, overhangs and large trees. Creating artificial shelters allowed flexibility in where they were located. Clothing enhanced their ability to withstand cold.

Cooperative hunting
Working in organised groups requiring considerable coordination, early humans were able to tackle large game that would be impossible for a solitary hunter.

1. Explain what is meant by **cultural evolution**: _____

_____

_____

Cultural Evolution

Code: RA 2

Development of Agriculture

People learned to plant and look after food plants, especially grains, and to domesticate animals. In the Middle East, about 8000 BC they learned to grow wheat, while in Mexico, about 500 BC they began to grow maize.

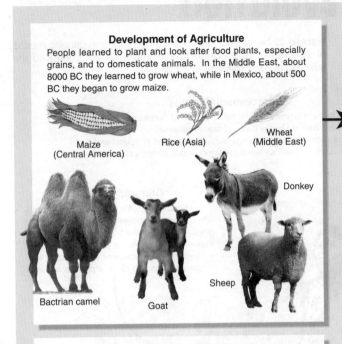

Maize
(Central America)

Rice (Asia)

Wheat
(Middle East)

Donkey

Bactrian camel

Goat

Sheep

Development of Stable Settlements

Communities of successful grain cultivators grew up, living in permanent, stable settlements of quite large size. Such people developed qualities such as patience, industry and a sense of property, and prepared the way for the next step.

Development of Cities

As communities became larger, trade and commerce began to develop. Large cities grew up where markets and trading systems developed. These were places where people could develop special skills such as pottery and metal work. It also resulted in rivalry between states and in wars.

The Knowledge Explosion

The sharing of ideas, and more free time for some in the cities resulted in a great speeding up of cultural evolution. In the last 200 years there has been a very rapid development of science and technology. Humankind developed the power to dominate the environment completely. In particular, medical science has largely solved the problems of infectious disease, and technology has allowed us to produce material wealth at a staggering rate.

The Present and the Future

Humankind's success has given us the problems of pollution and over-population. Not only can we modify our environment, but because of our knowledge of genetics, we are actually directing the evolution of other living things by selective breeding. It even seems likely that we may soon be able to direct our own evolution by actually altering genes. In fact we are reaching a stage where we have so much power to alter the environment that we need to think and act very carefully.

The move from opportunist scavenger to hunter-gatherer was a major stage in mankind's cultural evolution. It was taken in a series of small steps, over a very long time (perhaps a million years). A few human societies, such as the Australian aborigines last century, were still at this stage until very recently.

2. Describe two probable effects of a drying climate on the selection pressures directing the evolution of early hominins:

_____

_____

_____

3. Explain how each of the cultural developments listed below enhanced the survival ability of early humans:

(a) Manufacture of bone and stone tools: _____

_____

(b) Shelters and clothing: _____

_____

(c) Use of fire: _____

_____

(d) Cooperative hunting: _____

_____

(e) Development of agriculture: _____

_____

(f) Commerce and communication: _____

_____

Palaeolithic Tool Cultures

The **Palaeolithic (Old Stone Age)** refers to a time period in early human cultural development. It spans the emergence of the first recognisable stone tools about 2.6 million years ago in eastern Africa, until the development of sophisticated tool kits in the Mesolithic (Middle Stone Age) about 10 000 years ago. These tool cultures are known mostly by their stone implements. This does not mean that the associated hominins did not use other materials (such as wood), just that they did not preserve well.

Timeline of Stone Tool Technologies

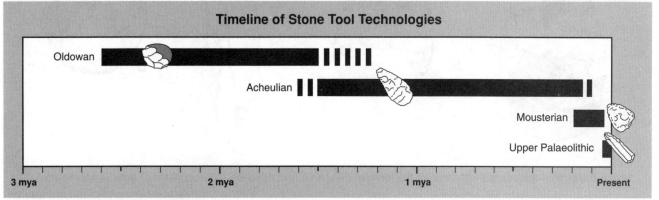

Oldowan

Acheulian

Mousterian

Upper Palaeolithic

3 mya   2 mya   1 mya   Present

Oldowan (pebble) tool culture

Probably made by *Homo habilis*, these tools were simple river-worn pebbles that were crudely fashioned with a minimum of flakes being removed. These tools typically had flakes knocked from several angles to produce a core with a cutting edge (e.g. chopper, discoid, polyhedron). Although the cores may have been used as tools, it is known that the sharp flakes were also useful in cutting.

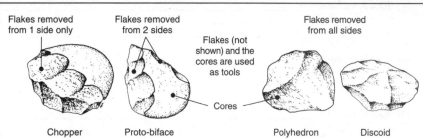

Flakes removed from 1 side only

Flakes removed from 2 sides

Flakes (not shown) and the cores are used as tools

Flakes removed from all sides

Cores

Chopper   Proto-biface   Polyhedron   Discoid

Acheulian tool culture

The product of *Homo erectus* and archaic *Homo sapiens*, these tools were typically 'tear drop' in shape and were carefully crafted with a slight bulge on each broad surface (called a bi-face). They ranged greatly in their size and are often referred to as hand axes although it is not clearly understood how they were used. They differ markedly from the earlier pebble tools in that there appears to be a standard design and each tool is manufactured using a great many more blows to remove flakes.

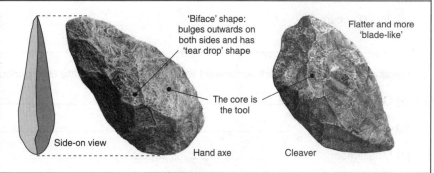

'Biface' shape: bulges outwards on both sides and has 'tear drop' shape

Flatter and more 'blade-like'

The core is the tool

Side-on view   Hand axe   Cleaver

Mousterian tool culture

The **Neanderthals** developed a more refined tool culture than the earlier Acheulian. Flint finally became a sought after material to produce stone tools. The advantage of this rock was the very predictable way in which it would chip when struck with another hard object. Much finer workmanship was possible. A particular tool making technique from this period is known as the **Levallois method**. It involves the preparation of a core and striking off a large oval flake which is then retouched on one surface only (see the photograph on the right; the retouched surface is visible).

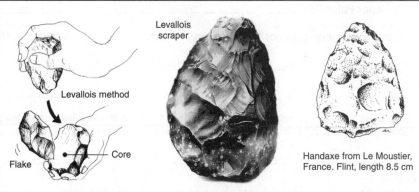

Levallois scraper

Levallois method

Flake   Core

Handaxe from Le Moustier, France. Flint, length 8.5 cm

Upper Palaeolithic tool culture

There was a rather sudden increase in the sophistication of tool making about 35 000 to 40 000 years ago. Both the **modern** *Homo sapiens* and the last of the Neanderthals produced flint tools of much finer workmanship using a technique called **punch blade**. Long, thin flakes are removed and shaped into a large number of different tool types. European sub-cultures (traditions) include the **Magdelanian**, **Solutrean** and **Aurignacian**. Other material such as bone, ivory and antler became increasingly utilised to produce very fine tools such as needles.

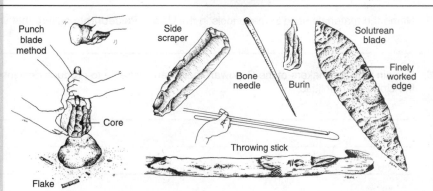

Punch blade method

Core

Flake

Side scraper

Bone needle   Burin

Solutrean blade

Finely worked edge

Throwing stick

Cultural Evolution

Code: A 2

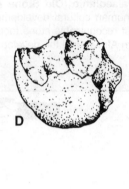

A

B

C

D

1. Name the culture associated with each of the tools above (A-D) and describe the features that help identify them:

(a) Tool **A** culture: _____

_____

(b) Tool **B** culture: _____

_____

(c) Tool **C** culture: _____

_____

(d) Tool **D** culture: _____

_____

2. Identify the **hominin species** associated with, and the approximate time period for, each of the tool cultures below:

(a) Oldowan: _____

(b) Acheulian: _____

(c) Mousterian: _____

(d) Upper Palaeolithic: _____

3. Describe the general trends in the design of the stone tool from **Oldowan** to **Upper Palaeolithic** cultures:

_____

_____

4. The tools that are recovered from early human prehistoric sites are almost invariably stone, bone or ivory. Explain why tools made from other materials are almost never recovered from these sites:

_____

_____

5. Name the materials used to make tools in the **Upper Palaeolithic** culture that were seldom used in earlier cultures:

_____

6. Explain why the makers of the **Oldowan** tools were not able to produce designs that were as sophisticated as those of the Upper Palaeolithic:

_____

_____

_____

_____

Palaeolithic Tool Use

It is impossible to tell with certainty what a tool recovered from an archaeological site was used for. By studying how similar tools have been used by recent 'stone age' societies, it is possible to infer their likely function. People using only stone-based technology were still in existence well into the first half of this century. Hunter-gatherer people existed in places like the Kalahari desert in south west Africa, the Australian outback, and some of the more remote areas of South East Asia and South America. Anthropologists studying these primitive cultures gathered valuable insights into how our ancestors may have lived.

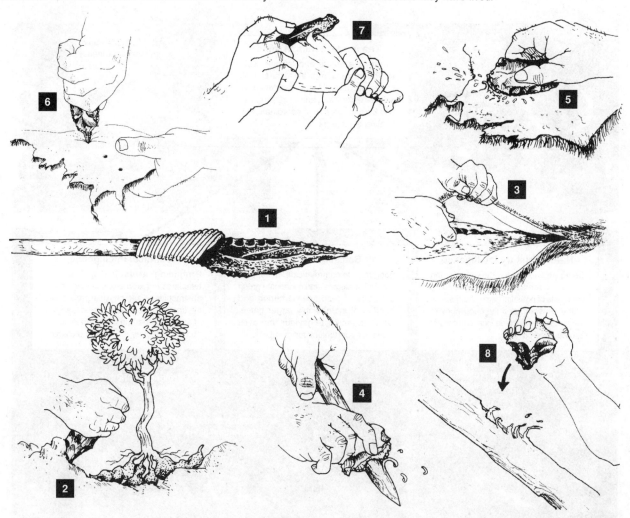

1. Match each of the diagrams above with the **description** of their function in the table below (place 1-8 in number column):

	No.	Name of Tool	Description
(a)			This tool was probably used to scrape the fat and sinew from the underside of a freshly killed animal, in preparation for curing.
(b)			The curved, sharp edge of this tool would probably have been used to shave wood chips from spear heads.
(c)			Later forms of this tool were probably used to skin and dismember carcasses.
(d)			A simple tool used as an axe, probably to cut into wood or possibly to dismember a carcass.
(e)			Early forms of this tool may have been used as a pick-like tool to expose root tubers growing under plants.
(f)			Used as a knife, this tool had only one side with a sharp cutting edge so that pressure could be applied to the blunt edge.
(g)			Hafted to a pole with greased sinew or plant fibre, this flint tool would have provided an effective cutting edge for the spear.
(h)			The sharp narrow point of this tool makes it an effective drill when twisted back and forth. In this way, holes could be made in materials such as hides, wood and ivory.

2. Assign each tool with their **correct** name from the list below:
    Side scraper, borer, denticulate tool, spear point, chopper, backed flake, early hand-axe, late hand-axe

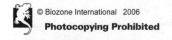

Cultural Evolution

Code: RA 2

Social Development

The evolutionary development of the brain in hominins was associated with increasing social complexity and the development of what we know as culture. Culture can be thought of as the accumulation of knowledge, rules, standards, skills, and mental abilities that humans utilise in order to survive. Obtaining and preparing food, coping with climatic conditions, trying to understand the world, and cooperating with others are normal, daily activities that require cultural solutions. The modern human mind is the cumulative product of responses to these various selection pressures.

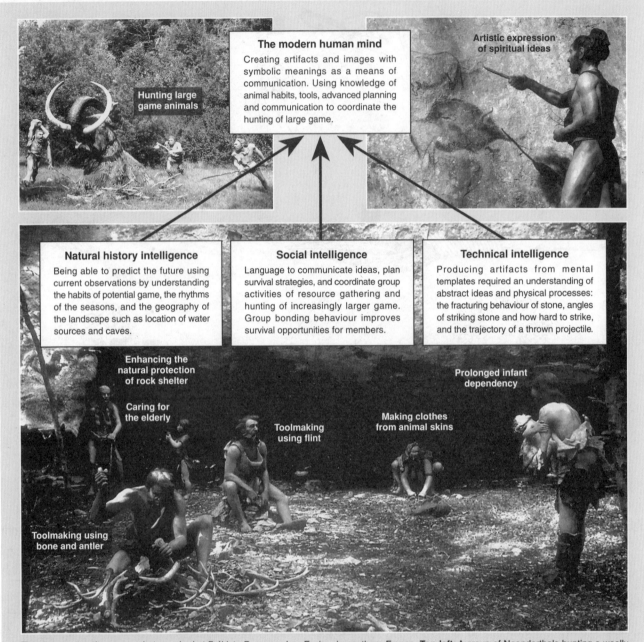

These reconstructions were photographed at Préhisto-Parc near Les Eyzies, in southern France. **Top left**: A group of Neanderthals hunting a woolly mammoth by using a concealed pit. **Top right**: A Cro-Magnon artist at work. **Bottom**: A Cro-Magnon family group at a rock shelter campsite. PHOTOS: RA

1. Explain how each of the following areas of human intelligence may have improved the success of early humans:

(a) Natural history intelligence: _____

(b) Social intelligence: _____

(c) Technical intelligence: _____

Art and Spirituality

Until recently, it was believed that art and spiritual beliefs first developed with the arrival of modern humans, particularly in Europe. A cultural explosion took place in the form of prehistoric art and new kinds of tools attributed to modern humans about 35 000 years ago. The beginning of this period, the **Aurignacian**, marks a dramatic development occurring simultaneously over large parts of western and eastern Europe. There is growing evidence that Neanderthals were more sophisticated than previously thought. Not only did they bury their dead, but they

appear to have adorned their bodies, used necklaces, and made and used flutes. The stimulus for the new cultural development, in the Aurignacian at least, was probably a need to represent, in a concrete and lasting way, ideas about the unknown, such as death, hunting success, and fertility. A wide range of materials were used in this display of early human thought. Ivory, bone, clay, and stone were used to create sculptures, and the walls of rock shelters and caves were adorned with drawings, paintings, and bass-relief (sculptures that stand out from the rock wall).

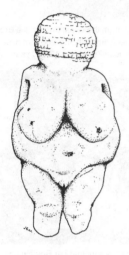

Sculpture: One of the earliest prehistoric works of art is the Venus of Willendorf (left), from Austria (10 cm high). The small figurine, carved from limestone about 30 000 years ago, has exaggerated breasts, buttocks, and body fat; these were thought to be desirable traits among women to enhance fertility and survival in periods of deprivation.

Music: Made of bird-bone, this 25 000 year old flute (above) is one of the earliest instruments providing evidence of music. A similar flutelike piece of cave bear bone has been found at a Neanderthal hunting camp in Slovenia. The bone, dated at between 43 000 and 82 000 years ago, suggests that Neanderthals may have made music.

Rock art: Many examples of rock art are found in southern France and northern Spain. However, rock art was not restricted to Europe; the earliest examples may come from Australia. The pigments (charcoal, manganese oxide, and ochre) were applied to grease smeared on the rock surface.

Animal bones, red ochre, flowers and horns buried with the body.

Position and orientation of the body with legs pulled up as if sleeping.

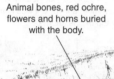

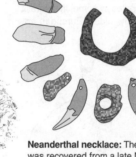

Neanderthal necklace: The necklace above was recovered from a late Neanderthal site. Was the necklace made by the Neanderthal, or was it traded or stolen from modern humans? Whatever the answer, it suggests that Neanderthals appreciated such objects.

Neanderthal burials: The Neanderthals of Europe and Southwest Asia buried their dead in a way that showed signs of ritualisation. The grave is usually characterised by certain items found buried with the body. The position and orientation of the body are consistently the same. Some experts have questioned the validity of these so-called burials and claim the finds are the result of coincidences and later disturbances.

Magic and ritual: Painted about 15 000 years ago on a cave wall at Trois Féres, in the Pyrenees Mountains, this creature (above right) may have been a sorcerer dressed in animal skins. Rituals involving spiritual re-enactment may have been conducted to ensure successful hunting.

1. (a) Identify three items that may have been found with the body in a Neanderthal grave: _____

   _____

   (b) Describe the usual position and orientation of skeletons found at Neanderthal burial sites: _____

   _____

   (c) Describe a cultural significance of burying the dead in this manner: _____

   _____

   _____

2. Describe two recent findings that support a view that Neanderthals were more sophisticated than previously thought:

   (a) _____

   (b) _____

Cultural Evolution

Code: A 2

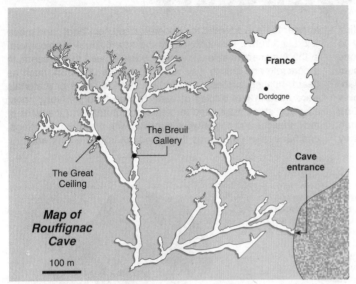

Detail of the head of a rhinoceros from the Great Frise in the **Breuil Gallery**, Rouffignac Cave. Like the mammoth, the woolly rhinoceros was common to Europe before becoming extinct 12 000 years ago.

The **Great Ceiling** is the heart of the sanctuary at Rouffignac. The collection of drawings depicts a mammoth and ibex. It was common for drawings to be made over the top of one another.

Rouffignac cave, in the Dordogne Valley in southwestern France, is a huge cavern with 10 km of galleries. It is remarkable for its large number of prehistoric paintings dating from the mid to late Magdelanian Period (~11 000 BC). The first drawings are about 750 m from the entrance and continue deep underground, to the end of the galleries. It is interesting to note that over 150 paintings of the mammoth in this cave represent half the known paintings of this animal. Early human inhabitants must have competed with cave bears for cave occupancy. Early humans often created their paintings in the most remote and inaccessible parts of the cave; an argument used in favour of a religion centred on the hidden forces of the Earth. Below and right are some typical examples of these prehistoric drawings. Earlier rock engravings have been dated (disputed by some) in Australia at >60 000 years; this is more than twice the age of the oldest European art.

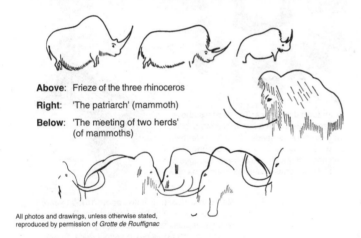

Above: Frieze of the three rhinoceros

Right: 'The patriarch' (mammoth)

Below: 'The meeting of two herds' (of mammoths)

All photos and drawings, unless otherwise stated, reproduced by permission of *Grotte de Rouffignac*

This painting of a horse is one of many splendid images painted on the ceiling of the famous Lascaux cave in the Dordogne, France. Note the abstract markings (arrowed) for which there is no explanation.

3. Describe the relationship the early humans had with the mammoth and woolly rhinoceros: _____

_____

_____

4. Describe a probable purpose for the drawings and paintings appearing in the cave art: _____

_____

_____

_____

5. Describe the likely function of the 'Venus' statues in Palaeolithic life (see the *Venus of Willendorf* on the previous page):

_____

_____

_____

Mesolithic and Neolithic Cultures

Following the Upper Palaeolithic period, which ended about 15 000 to 10 000 years BC, new cultures arose. About 15 000 years B.C., Kebaran hunters (in Palestine) included wild cereal grasses into their diet. The **Mesolithic** (Middle Stone Age) culture lasted for a few thousand years, from about 12 000-10 000 years ago. This culture is characterised by the use of small stone tools (microliths) and by a broad-based hunter/gatherer economy after the last ice age. It is believed to have involved foraging for seeds from wild cereal grasses. The **Neolithic** (New Stone Age) culture marked a major turning point in the procurement of food. This period is usually associated with the beginnings of agriculture, pottery, and permanent settlements in the Old

World. Farming began in parts of western Asia the so-called 'fertile crescent' running from Egypt to the Persian Gulf about 10 000 years ago (although without pottery). By about 7000 years ago, agriculture became established in China, followed by Mesoamerica (Guatemala, Honduras and southern Mexico) about 5000 years ago. The significance of this shift away from a hunter-gatherer economy to one which could provide surplus food, was far reaching. It meant that greater population densities could be achieved and not all individuals needed to be involved in food gathering activities. This allowed for the development of artisans and accelerated the cultural development of an expanding human population.

The Origin of Agriculture

Mesoamerica — Beans, maize, peppers, squash, gourds, cotton, guinea-pigs, llamas — 5000 years ago

Fertile Crescent — Barley, wheat, Emmer, Einkorn, lentils, peas, sheep, goats, cattle — 10 000 years ago

North China — Rice and millet — 7000 years ago

Southest Asia — Rice, bananas, sugar cane, citrus fruits, coconuts, soya beans, yam, millet, tea, taro, pigs

South America — Lima beans, potatoes, squash, beans, and pumpkins

Africa — Millets, sorghum, groundnuts, yams, dates, coffee, and melons

Maize

Each domesticated animal was bred from the wild ancestor. The date indicates the earliest record of the domesticated form.

Domesticated animal	Wild ancestor	Region of origin	Date (years ago)
Dog	Wolf	many places?	13 000
Goat	Bezoar goat	Iraq	10 000
Sheep	Asiatic mouflon	Iran, Iraq, Levant	10 000
Cattle	Aurochs	Southwest Asia	10 000
Pig	Boar	Anatolia	9000
Domestic fowl	Red jungle fowl	Indus Valley	4000

Domesticated animal	Wild ancestor	Region of origin	Date (years ago)
Horse	Wild horse	Southern Ukraine	6000
Arabian camel	Wild camel	Southern Arabia	5000
Bactrian camel	Wild camel	Iran	4500
Llama	Guanaco	Andean plateau	6000
Water buffalo	Indian wild buffalo	Indus Valley	4500
Ass	Wild ass	Northeast Africa	5500

1. The **Mesolithic culture** replaced the Upper Palaeolithic culture.

(a) State when the Mesolithic culture began: _____

(b) Explain how it differed from the Upper Palaeolithic culture: _____

2. The **Neolithic culture** replaced the Mesolithic culture.

(a) State when the Neolithic culture began: _____

(b) Describe the important cultural developments of the Neolithic culture: _____

Cultural Evolution

Code: A 2

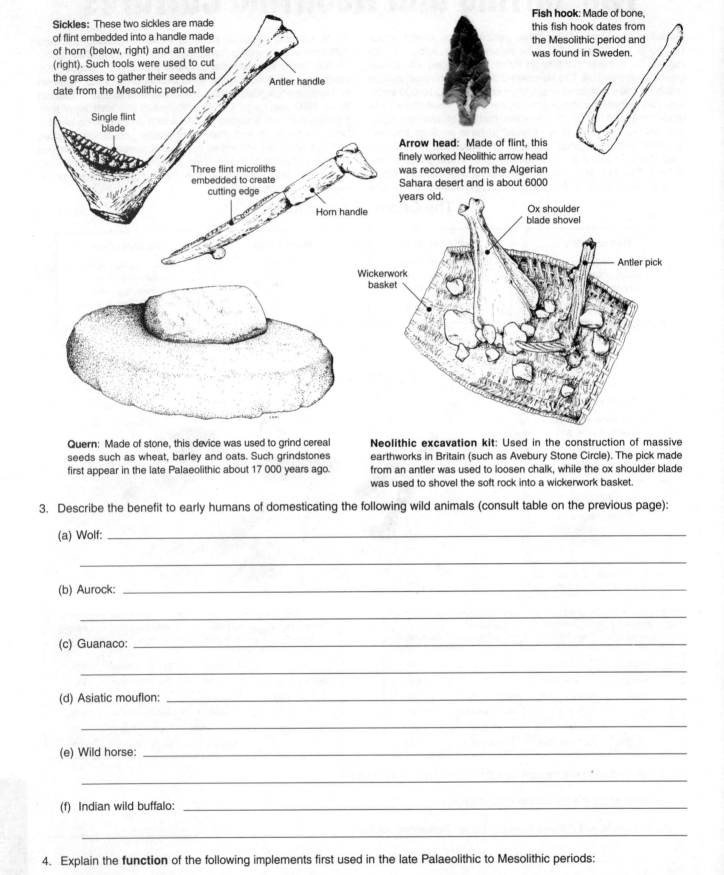

Sickles: These two sickles are made of flint embedded into a handle made of horn (below, right) and an antler (right). Such tools were used to cut the grasses to gather their seeds and date from the Mesolithic period.

Antler handle

Single flint blade

Three flint microliths embedded to create cutting edge

Horn handle

Fish hook: Made of bone, this fish hook dates from the Mesolithic period and was found in Sweden.

Arrow head: Made of flint, this finely worked Neolithic arrow head was recovered from the Algerian Sahara desert and is about 6000 years old.

Ox shoulder blade shovel

Antler pick

Wickerwork basket

Quern: Made of stone, this device was used to grind cereal seeds such as wheat, barley and oats. Such grindstones first appear in the late Palaeolithic about 17 000 years ago.

Neolithic excavation kit: Used in the construction of massive earthworks in Britain (such as Avebury Stone Circle). The pick made from an antler was used to loosen chalk, while the ox shoulder blade was used to shovel the soft rock into a wickerwork basket.

3. Describe the benefit to early humans of domesticating the following wild animals (consult table on the previous page):

(a) Wolf: _____

_____

(b) Aurock: _____

_____

(c) Guanaco: _____

_____

(d) Asiatic mouflon: _____

_____

(e) Wild horse: _____

_____

(f) Indian wild buffalo: _____

_____

4. Explain the **function** of the following implements first used in the late Palaeolithic to Mesolithic periods:

(a) Quern: _____

_____

(b) Sickle: _____

_____

_____

Bronze Age Culture

Copper was in use as early as perhaps 9000 years ago in Southeast Turkey. The beginning of this phase in cultural development is sometimes referred to as the Chalcolithic (Copper-Stone) Age. This is because at the same time, stone technology was still very much in use. Copper was initially very scarce and was only used for small and precious objects such as beads and pins. This early stage only involved the working of native copper metal by cold-hammering without smelting. Furnaces were used to smelt the ore of copper about 6600 years ago in a large area that spans south-western Europe to south-western Asia. Metalsmiths learned to pour molten copper into stone moulds and as skills in metallurgy rapidly developed, tools and weapons were cast in copper. This was a major factor in the urbanisation of Mesopotamia. The practice then spread into the Mediterranean area and infiltrated the Neolithic cultures of Europe. True bronze, an alloy of tin and copper, did not appear until 5000 years ago. Tin deposits were hard to find and it was not until about 4000 years ago that more considerable deposits were located and the use of bronze flourished. For almost 3000 years, copper and then bronze, were the main metals used for making implements, vessels and weapons. Stone continued to be used in decreasing amounts for these purposes as the availability and use of metals increased. About 3000 years ago, the forging of iron brought the Bronze Age to an end.

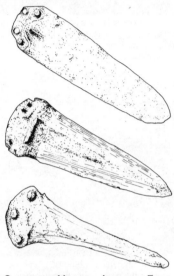

Copper and bronze daggers: From early Bronze Age sites around Stonehenge and Avebury in southern England, these blades would have had wooden or bone handles attached.

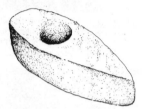

Axe head: Made of stone, this axe head would have been hafted to a wooden handle.

Jet button: Made of a hard black variety of lignite, jet can be given a brilliant polish. It was commonly used for items of jewelry such as this button.

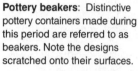

Pottery beakers: Distinctive pottery containers made during this period are referred to as beakers. Note the designs scratched onto their surfaces.

Flint knife: Although copper and bronze became increasingly used to make tools and weapons, they were still used in conjunction with traditional materials such as flint.

1. Describe what the Chalcolithic Age was and when it began: _____

_____

2. Explain the limitations of copper as a metal for making implements: _____

_____

3. Describe an advantage that bronze had over previous traditional materials for making implements: _____

_____

4. Explain what is meant by native copper and how it was originally worked before furnaces were invented: _____

_____

_____

5. Describe the composition of 'true bronze': _____

6. State when the Bronze Age began and when it started to be displaced by iron: _____

_____

7. Explain why so little bronze was made before 4000 years ago: _____

_____

8. Identify materials other than copper or bronze that were used as implements during this period: _____

_____

_____

Code: A 1

Recent Cultural Evolution

Year BC	Cultural Period

Mesolithic

9000	Jericho settled as first town
	Sheep domesticated in Middle East
	Tools fashioned from native copper in use (Anatolia)

Neolithic

8200	Goat domesticated in Iran
8000	Pottery made in Japan
7500	Humans cultivate the first crops (wheat and barley) in the Middle East
7200	Sheep, pigs, cattle (domesticated) near Greece
6900	Domesticated dog at Sarab (Iran)
5500	Open mould copper casting at Turkey
	Copper used as trade in Mediterranean area

Copper Age

5000	Farming supersedes hunting in most of Europe
	Corn grown in Mexico
	Vaulted graves in Brittany
4350	Horse herding in the Ukraine
4000	European metallurgy. (Copper axe casting in Rumania.)
	Sail boats used in Europe
	Large cities developed
3500	Writing appears. Many trade routes evident
	Egyptian merchant ships ply Mediterranean
3120	Beginning of Dynasty 1 Egypt
3100	Goldsmiths active in Bulgaria

Bronze Age

3000	Tin bronze implements found at royal cemetery 'Ur'
2800	Calendar devised in Egypt
2613	Beginning of Dynasty IV in Egypt. The Pyramid Era
2300	Very large cities in existence - Ebla - 260 000 people
	Trade flourishes via the Persian Gulf
1900	Pharaoh Sesostris II builds the Fayum reservoir
1780's	Reign of Hammurabi in Mesopatamia during which many of the major law codes controlling Near East trade were introduced
1595	End of first Dynasty in Babylon
1500	Invention of ocean going outrigger canoes enables man to reach the islands of the South Pacific
1480	Paved roads in Crete

Iron Age

1400	Iron in use in Middle East
1350	Alphabet devised in Syria
1300	Turin Royal Papyrus - records the first dynasties in Egypt
1000	Reindeer domesticated in Northern Europe
900	Phoenicans develop modern alphabet
800	Celts spread iron through Europe
	Nomads build a society based on the horse in Russia
720	Homer composes Iliad and Odyssey
700	Rome founded
	Civilisation flourishes and expands
0	Christian era begins

The stone circle of Avebury is one of the most famous of the many Neolithic stone circles in Britain. Built in stages during the late Neolithic period 2500-2000 BC, its purpose is unknown, but was probably a place of worship. The 150 000 t of chalk excavated during its construction indicates its importance.

Egyptian culture bloomed during the 4th dynasty (2500 BC). This was a wealthy, stable period that ushered in the civilisation's first great age, the Old Kingdom. The Giza pyramids and the sphinx (above), which bears the face of king Khafre, are unparalleled feats of architecture of the era.

A reconstruction of a European Iron Age (c. 500 BC) dwelling: a round house, located at the Butser Hill Experimental Iron Age Farm in Southern England. The building would have been constructed from wicker-work (woven branches) with a thatched 8m high roof.

The Mayas of central America were the only truly literate civilisation of the Americas, dating back to 300 BC. Large architectural complexes formed the centres of cities with up to 50 000 inhabitants. Plazas surrounded by stone pyramids were crowned by palaces and temples for human sacrifice.

Present and Future Human Evolution

Modern humans encounter selection pressures that are quite different to those experienced by our early ancestors. No longer do we have to escape or fight predators, track down and subdue game animals, or follow migratory herds over kilometres of harsh terrain. Some of the trends identified in the evolution of human anatomy (described below) may continue, but new selection pressures, arising as a result of our modern culture and technology, may increasingly influence our evolution. Pollution of our water, soil, and air introduces toxic substances into our food, and our ability to tolerate and metabolise these substances may become increasingly important. Increased penetration of UV radiation, as a result of depleted stratospheric ozone, may have already led to an increase in the incidence of skin cancers. Individuals with a high tolerance to electromagnetic radiation from electrical devices such as cellular phones may be favoured in our modern environment. The development of **genetic engineering** may enable us to alter our genetic makeup and reduce our susceptibility to detectable disorders. Increasingly, medical advances enable people with inherited, potentially lethal, disorders to live long enough to bear children. Increased population mobility increases gene flow between previously isolated populations, and has seen the emergence and spread of new diseases. The future evolution of humans will be determined by our responses to these new selection pressures.

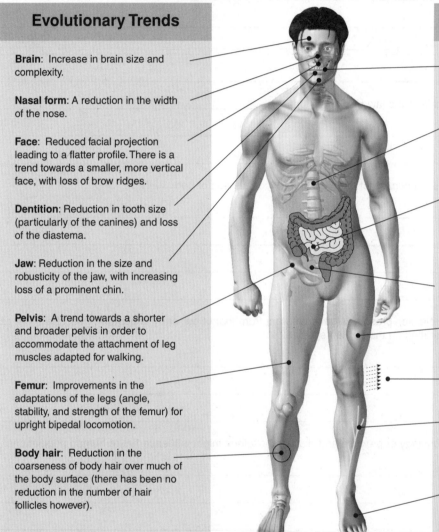

Evolutionary Trends

Brain: Increase in brain size and complexity.

Nasal form: A reduction in the width of the nose.

Face: Reduced facial projection leading to a flatter profile. There is a trend towards a smaller, more vertical face, with loss of brow ridges.

Dentition: Reduction in tooth size (particularly of the canines) and loss of the diastema.

Jaw: Reduction in the size and robusticity of the jaw, with increasing loss of a prominent chin.

Pelvis: A trend towards a shorter and broader pelvis in order to accommodate the attachment of leg muscles adapted for walking.

Femur: Improvements in the adaptations of the legs (angle, stability, and strength of the femur) for upright bipedal locomotion.

Body hair: Reduction in the coarseness of body hair over much of the body surface (there has been no reduction in the number of hair follicles however).

After C.B. Stringer

Associated Problems

Teeth: Overcrowding and impaction of wisdom teeth as a result of reduction in the size of the jaw (a recent development in evolution).

Slipped disc: Lower back troubles, usually the result of degenerative changes with age, are compounded by the load being carried by only two limbs.

Hernias: The intestines may bulge out through our weakened abdominal wall. This is the result of the gut no longer being hung from the spine by a broad ligament as it is in quadrupeds; the problem is compounded by obesity.

Birth canal (in women): Changes in pelvic shape in response to bipedalism, together with babies born with larger skulls, cause childbirth problems.

Skin cancer: With the loss of shielding body hair, the skin is exposed to larger amounts of UV radiation from the sun.

Hypothermia: Excessive loss of heat as a result of reduced body hair becomes a problem in cooler latitudes.

Varicose veins: An upright posture hampers venous return, allowing blood to collect in the leg veins. Blood must overcome about 1.2 m of gravitational pressure to return to the heart.

Flat feet: Feet may suffer strain because the body rests on just two limbs. The arches of the feet collapse resulting in flat footedness, distorted bones, hammer toes, and bunions.

1. Describe the evolutionary trends associated with the following human anatomical features:

(a) Body hair: _____

_____

(b) Jaw: _____

_____

(c) Brain: _____

_____

(d) Pelvis: _____

_____

Cultural Evolution

Code: RA 2

72

(e) Face: _____

_____

(f) Femur: _____

2. Describe three problems with the modern human anatomy that have developed as a result of bipedalism:

(a) _____

_____

(b) _____

_____

(c) _____

_____

3. Describe a problem that has resulted from the jaw becoming smaller: _____

_____

_____

4. Describe two problems that have arisen because of the reduction of coarseness of body hair:

(a) _____

_____

(b) _____

_____

5. Future human populations will face selection pressures quite different from that of our early ancestors. Describe some of the selection pressures currently operating on the human species:

_____

_____

_____

6. Describe examples of a modern technology or practice and discuss how they might influence future human populations:

_____

_____

_____

_____

_____

_____

_____

_____

_____

_____

_____

_____

Trends in Human Evolution

Use the information you have collected during the study of this topic to fill in the summary sheet for each of the hominins listed below. Some points to consider are listed on the right. Not all boxes can be filled in.

Skull
Consider such things as:
1. Skull shape (e.g. brow ridges).
2. Brain capacity & organisation (which lobes are prominent?).
3. Face angle: snout present?
4. Tooth size, enamel, and use.
5. Attachment of skull muscles.
6. Robustness of skull.

Locomotion
Consider such things as:
1. Bipedal or quadrupedal?
2. Position of foramen magnum.
3. Knee joint and femur shape.
4. Valgus (carrying) angle.
5. Foot structure.
6. Lower spine shape.
7. Pelvis and hip joint.

Tool use & manufacture
Consider such things as:
1. Tool technology: Oldowan, Acheulian, Mousterian, upper Palaeolithic (how finely worked). Purpose of tools.
2. Other materials used: antler, wood, bone, pottery, metals.
3. Use of fire.

	Skull	Locomotion	Tool use & manufacture
Homo sapiens Duration: _____ Years ago			
Homo neanderthalensis Duration: _____ Years ago			
Archaic Homo sapiens Duration: _____ Years ago			
Homo erectus Duration: _____ Years ago			
Homo habilis Duration: _____ Years ago			
Australopithecus afarensis Duration: _____ Years ago			

Cultural Evolution

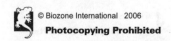
Code: RA 3

Abstract thought

Consider such things as:
1. Burial of dead.
2. Collecting materials for tool manufacture, complex design.
3. Artistic expression: paintings, statues, carvings.
4. Writing and recording events.

Diet and food resources

Consider such things as:
1. Vegetarian, carnivore or opportunist omnivore?
2. Food collection: scavenger, hunter/gatherer, small or large game, domestication of plants and animals.
3. Suitability of teeth for diet.

Distribution

Consider such things as:
1. Globally: Africa, Asia, Europe, Americas and the Pacific.
2. Habitat: savannah, forest, deserts, mountains, coastal.
3. Presence of other hominin species in the same place at the same time.

Selection pressures

Consider such things as:
1. Climatic change (e.g. ice ages).
2. Diet on: teeth, jaw, muscles.
3. Competition with other species.
4. Predator avoidance.
5. Effect of bipedal locomotion on other body parts.
6. Cultural response to pressures.

Abstract thought	Diet and food resources	Distribution	Selection pressures

Index